Simultaneity and Delay

Continuum Studies in Continental Philosophy
Series Editor: James Fieser, University of Tennessee at Martin, USA

Continuum Studies in Continental Philosophy is a major monograph series from Continuum. The series features first-class scholarly research monographs across the field of Continental philosophy. Each work makes a major contribution to the field of philosophical research.

Adorno's Concept of Life, Alastair Morgan

Badiou, Marion and St Paul, Adam Miller

Being and Number in Heidegger's Thought, Michael Roubach

Deleuze and Guattari, Fadi Abou-Rihan

Deleuze and the Genesis of Representation, Joe Hughes

Deleuze and the Unconscious, Christian Kerslake

Deleuze, Guattari and the Production of the New, edited by Simon O'Sullivan and Stephen Zepke

Derrida, Simon Morgan Wortham

Derrida and Disinterest, Sean Gaston

Derrida: Profanations, Patrick O'Connor

The Domestication of Derrida, Lorenzo Fabbri

Encountering Derrida, edited by Simon Morgan Wortham and Allison Weiner

Foucault's Heidegger, Timothy Rayner

Gadamer and the Question of the Divine, Walter Lammi

Heidegger and a Metaphysics of Feeling, Sharin N. Elkholy

Heidegger and Aristotle, Michael Bowler

Heidegger and Logic, Greg Shirley

Heidegger and Nietzsche, Louis P. Blond

Heidegger and Philosophical Atheology, Peter S. Dillard

Heidegger Beyond Deconstruction, Michael Lewis

Heidegger, Politics and Climate Change, Ruth Irwin

Heidegger's Early Philosophy, James Luchte

Idealism and Existentialism, Jon Stewart

Kant, Deleuze and Architectonics, Edward Willatt

Levinas and Camus, Tal Sessler

Merleau-Ponty's Phenomenology, Kirk M. Besmer

Nietzsche, Nihilism and the Philosophy of the Future, edited by Jeffrey Metzger

Nietzsche's Ethical Theory, Craig Dove

Nietzsche's Thus Spoke Zarathustra, edited by James Luchte

The Philosophy of Exaggeration, Alexander Garcia Düttmann

Time and Becoming in Nietzsche's Thought, Robin Small

Sartre's Phenomenology, David Reisman

Who's Afraid of Deleuze and Guattari? Gregg Lambert

Žižek and Heidegger, Thomas Brockelman

Žižek's Dialectics, Fabio Vighi

Simultaneity and Delay

A Dialectical Theory of Staggered Time

Jay Lampert

Continuum Studies in Continental Philosophy

continuum

Continuum International Publishing Group
The Tower Building
11 York Road
London SE1 7NX

80 Maiden Lane
Suite 704
New York NY 10038

www.continuumbooks.com

British Library Cataloguing-in-Publication Data
A catalogue record for this book is available from the British Library.

ISBN: HB: 978-1-4411-2639-9

Library of Congress Cataloging-in-Publication Data
Lampert, Jay, 1957–
Simultaneity and delay : a dialectical theory of staggered time/Jay Lampert.
p. cm. – (Continuum studies in Continental philosophy)
Includes bibliographical references (p.) and index.
ISBN 978-1-4411-2639-9 (hardcover)
1. Time. 2. Synchronization. I. Title.
BD638.L255 2011
115–dc23 2011037080

Typeset by Deanta Global Publishing Services, Chennai, India
Printed and bound in Great Britain

Contents

Preface ix

1. Introduction to Concepts 1
   (a) Simultaneity 2
   (b) Delay 12
2. Phenomenology of Simultaneity and Delay: The Problem of Converging Timelines 21
   (a) Simultaneity in Husserl 21
   (b) Delay in Husserl, Merleau-Ponty and Heidegger 30
3. Simultaneity and Delay in the Era of Eternity: Ancient and Medieval Issues 49
   (a) Simultaneity as Synchronization: Plato and Plotinus 49
   (b) Delay as Deferred Eternity: Plato and Origen 57
4. Simultaneity and Delay in the Era of Time: Late Modern Issues 75
   (a) Simultaneity in Kant: The Schema for Returning to the Same Place: Simultaneity as Reversible Succession 75
   (b) Simultaneity in Hegel: The System of Expanding Explications: Succession via Divisible Simultaneity 82
   (c) Delay in Lessing and Historical Dialectic: Lingering and Repetition 93
5. Simultaneity and Delay in the Sciences 97
   (a) Simultaneity Relative to Frames of Reference: Physics 97
   (b) Delay in, but not Observed by, Consciousness: Cognitive Science 112

6. Simultaneity and Delay in the Era of Simultaneity: Bergson 123

(a) Disjunctive Simultaneity as Virtual Substitution: *Duration and Simultaneity* 123

(b) Delay as Rhythmic Oscillation: *Matter and Memory* 137

7. Simultaneity and Delay in the Era of Delay 149

(a) Delay as *Différance*: Derrida's Solution to Two Problems of Time in Husserl 150

(b) Simultaneity as Pure Past: Deleuze's Solution to the Problem of Time in Bergson 159

8. Simultaneity and Delay in Politics 171

(a) Delay as Diachrony to Come: Balibar, Deleuze and Derrida 172

(b) Simultaneity and Decision: Badiou 189

9. Simultaneity and Delay in Music 197

(a) Simultaneity and Seriality: Schoenberg and Adorno 199

(b) Delay and the Unpulsed Time-Arrow: Boulez, Deleuze and Spectral Music 215

10. Conclusion 227

(a) Organizing the Simultaneity-Delay Timescape 227

Notes 233

Bibliography 249

Index 259

Preface

There should have been more about X in this book. All criticisms in this form are well founded. The philosophers most used in this book—Husserl, Bergson, Derrida and Deleuze—all require a great deal more analysis. Other philosophers discussed—Plato, Aristotle, Kant, Hegel, Merleau-Ponty, Heidegger, Balibar and Badiou—should have been used more. Philosophers only mentioned should have been discussed. Commentaries on all these philosophers should have been taken up. I should have become an expert on physics, cognitive science, and music theory. This book is too short.

I would like to thank the University of Guelph for generous support of research leaves and Chairs Andrew Bailey and Andrew Wayne. Many graduate students at the University of Guelph, where I presented some of this material for the first time in graduate courses, have been extraordinarily helpful in generating ideas and sending me things to read, particularly Jason Robinson and David Peck. Jill Gilbert did a brilliant job transcribing my lectures. Francis Coté was a great Undergraduate Research Assistant. Casey Ford, as a Graduate Research Assistant, helped me with final bibliographical notes. All my graduate students during this period have contributed enormously to my research.

Many people have been very supportive of this research: Len Lawlor, Costas Boundas, and Fredric Jameson in particular. I have benefited immeasurably (and dialectically) over the years from discussions of this material with Jennifer Bates. Karen Houle has given me all sorts of ideas. Earlier versions of portions of this book were presented at Marquette University, Villanova University, Duquesne University, and the University of Manitoba, as well as what used to be the joint Doctoral Program of Guelph/McMaster University/Wilfred Laurier University. Fred Evans provided helpful suggestions on a synoptic version. James Swindal at Duquesne University has contributed by his support. Philosophical conversations with Joshua Kates, Dan Selcer, Jim Vernon, and many other brilliant philosophers, have been indispensable. The Médiathèque at IRCAM in Paris is a wonderful resource.

And I would like to thank Jennifer for many things, and Hector, for his awesomeness.

Page numbers (and sometimes section numbers) in parentheses in the body of the manuscript refer to the most recently named author's text, as referenced in the most recent footnote.

Chapter 1

Introduction to Concepts

Every region of being presupposes some degree of synthesis, synchronization, or simultaneity. But total synchronization is impossible—it is delayed. Delay is not an epistemic aporia; it is the real structure of temporality. Simultaneity is always being actualized, but only in its delay-forms.

The problem of simultaneity—of what it means for events to happen at the same time—strikes philosophy in the early twentieth century, with Bergson and Husserl. In the late twentieth century, Derrida and Deleuze challenge the whole idea of 'the same time' and instead thematize delay—events staggered over distant times. This struggle between convergent and staggered time plays out also in twentieth-century aesthetics, politics, and the sciences.

The thesis of this book is that the moment where many things happen 'at the same time' is constructed out of converging rhythms and then unfolds in delayed reactions. Time becomes a shifting continuity of events at a distance. The dialectic of simultaneity and delay organizes time into elastic rhythms. With this model, we can analyze the time-structures of such diverse phenomena as atonal music, political decision-making, leaps of memory and the boredom of waiting, and simultaneities and delays in everyday experience and behaviour.

Simultaneity and delay are difficult notions. Simultaneity involves the convergence of two or more objects whose individual time-lines follow different rhythms. Delay describes the situation where events seem as though they should be happening at the same time, but one of them lags behind. Without an ideal of simultaneity, delay would have no reference. Simultaneity and delay presuppose one another, even as opposites. Time is a multi-levelled system that pulls towards both convergence and divergence.

The problems of simultaneity and delay have an impact on general problems in the philosophy of time: for example, on whether the dimension of time is analogous to dimensions of space; whether there is an irreversible time arrow; whether the present is specious and whether past and future exist at all; whether time is objective, subjective, or intersubjective; whether time has a beginning and end.

This book employs three methodologies. The first is to gather relevant phenomena from a wide range of sources, to organize related categories (like contemporaneity, coexistence, and synchronization; deferral, waiting, and anachronism), and to problematize the concepts of simultaneity and delay (both of which challenge the concept of succession). My intent is not to decide which phenomena are paradigmatic (e.g. whether the paradigm simultaneity is that of a visual field, or coexisting causes, or calculations in relativity physics, or political conventions like time zones). My intent is to weave relevant phenomena into an overall conceptual description.

The second method of this book is to wind a pathway through the many unsystematic analyses of simultaneity and delay sprinkled through the history of philosophy, as well as some in the arts and sciences. I emphasize Husserl and Bergson, Derrida and Deleuze, but philosophers from Plato and Plotinus to Kant and Hegel to Heidegger and McTaggart make important contributions to these concepts and leave behind further problems to solve. My intent is to uncover a conceptual order in this history.

The third method is to organize these results. It may be too optimistic to deduce theorems from principles, following Proclus, Spinoza, Hegel, or Badiou. Yet, events that include both same and different times generate a vast and intriguing range of simultaneity and delay phenomena. In the Conclusion, I offer a few notes towards a project of systematization.

There are many themes implicit in simultaneity and delay: speed and synchronization, desire and delayed gratification, revolution and anachronism[1], memory and fidelity, eschatology and return, intersubjective pairing, alternative histories, and the problem of the time-arrow. This network of concepts generates a complex timescape applicable to such phenomena as the illusion of movement in sculpture, the stacking of rhythms in music, flashbacks and cliffhangers in cinema, the power of indecision, the spectre of impending events that never come, political solutions that come too-little-too-late, time zones and time lags (dates and schedules), delayed reactions to trauma, squandered time, techniques for catching-up, hold-ups and slow-downs, as well as deceptively simple synchronizations in everyday life, like walking and eating.

(a) Simultaneity

Simultaneity covers 'happening at the same time', whether that time is in the past, present, or future. I allow myself to refer to 'simultaneous' events or to 'a simultaneity', that is, a network of simultaneously existing events.

One could argue that there is no such thing as pure simultaneity (that a temporal moment is not even identical with itself). Or one could argue that reality is eternal, all of whose parts exist simultaneously. One could ground simultaneity in time or in space, in passive experience or active construction. Simultaneity may be perceivable or inferable; it may relate real or virtual events. Sometimes simultaneity refers to two aspects of one process; sometimes to two interacting processes; sometimes to the coincidence or synchronicity of unrelated processes. Simultaneity may refer to events at the identical time or to overlapping, staggered, or commensurable events on a common measure. The 'same' in 'at the same time' raises all the logical and epistemological problems of 'sameness' in general.

Simultaneity seems to be not about things changing over time but about things that are just there at the same time. But I will emphasize that many temporal processes go into simultaneity and that moments of non-sameness are included 'at the same time'. Yet we should not conclude that simultaneity does not exist—we should rather be prepared for it to be complicated and peculiar. When we think of a time-line on which several events are located at the same point, we may at first imagine points on a simple straight line, like time-codes on filmstrip. But historians over the centuries have represented time-lines in amazing ways. It is difficult to line up independent rhythms on a time-line. Year-by-year chronologies do not easily map onto genealogies of kings who live for different spans. And it is difficult to tabulate dates from different calendars, for example, speculative Biblical dates against confirmed Roman dates. This leads drafters of time-lines to experiment with almost every imaginable graphic design. A survey of the history of time-lines reveals time-'lines' pictured as time-trees, time-rivers, concentric time-wheels, kaballic time-bodies, frenzied gothic lines, time-mazes, time-atlases, time-hydrolics, time-cartoons, time-palaces, Doomsday clocks and astrological convergence-charts, palm-lines readied for chronological palm-reading, immense fold-out 'synchronologies', and time-line graphics that look more like piles of hair or over-wired computer chips.[2]

The phenomena

To piece together the concept of simultaneity, we want to collect the full range of phenomena. The Socratic problem is to know which phenomena to gather together if we do not already have a definition. We might think of gathering phenomena in a Heideggerian sense, bringing diverse objects into a focal point of consciousness, constructing a world of potentiality.

To begin, we have to ask whether time is represented (a) in instants or (b) only in durations. Indeed, (c) time may not pass at all, and (d) it may have to be constructed by social or conceptual acts.

(a) Instantaneous simultaneities may take the form of (i) two subjective experiences at the same time or (ii) two objects or events at the same time.

(i) The paradigm of instantaneous subjective simultaneity involves phenomena sharing a visual field. Things in space can be present as a totality independent of the temporal order in which we examine them. Simultaneity here is more a spatial concept, describing objects laid out alongside one another, than a temporal concept involving succession.

There are other paradigm simultaneities in experience besides space, like the simultaneous experience of shape and colour. In fact, perceiving colours simultaneously is already complicated. Michel-Eugène Chevreul's 'Law of Simultaneous Contrast' (1839) shows that the intensity, hue, temperature, and chroma (brightness) of a perceived colour depends on the brightness, complementarity, and size of the thing adjacent to it simultaneously. For example, the same green square looks brighter if surrounded by a bright yellow square than by a dark magenta square. Two colours ('metamers') may match under one light source but not under another.[3]

There are also logical simultaneities, as when a person simultaneously believes one proposition, and without being conscious of doing so, believes its implications. It is not easy to say what experiential simultaneity in all these cases consists in. There may be structures and Gestalts that contain a multiplicity of experiential contents. There may be acts that have to be performed in order to collect contents into one simultaneous picture. Alternatively, simultaneity may be something of an illusion. When I notice something out of the corner of my eye, I can redirect my attention to it so quickly that I seem to see it at the same time. But it does not seem right to reduce simultaneity either to atemporal structure or to rapid succession.

For Gadamer, objective simultaneity (in his terms, *Simultaneität)* is less interesting than its subjective recognition (which he calls 'contemporaneity', *Gleichzeitigkeit*): tasting artworks of past ages, for example, is more interesting than the objective fact that historical works survive (damaged[4]) into the present.[5]

(ii) Nevertheless, we can chart objective simultaneity in all sorts of phenomena by matching multiple spatial coordinates against shared points on the time-axis.

(b) But of course, the idea of a temporal instant or point is not very plausible, except artificially. If we reject instants, then instead of saying two events occur at the same time, we should say they occur over the same duration. (Bergson prefers to call this 'contemporaneity'.) But if there are no instants at which a duration begins and ends, we have to speak of simultaneity in terms of overlapping staggered events, fading indefinitely in and out of synch. This makes simultaneity, whether experiential or objective, difficult to pinpoint.

(c) There is a view that time-passage is a myth. For McTaggart, the past-present-future relation is illusory, or at best merely subjective (see Chapter 5). McTaggart concluded that time as such is an illusion, but some of his followers conclude that time is not about flow or passage but about before-after relations. The objective fact that Paul Henderson shot a puck at t^1 before it entered the net at t^2 is indifferent to what is present now. Like all temporal facts, it is true at all times. Most anti-time-flow philosophers do not think that time is *totum simul*, a block of changeless being, but only that the totality of before-after relations permanently constitutes the time-order of the world. Still, there is something of total simultaneity in the view that before-after relations are permanent fixtures of the time-axis of space-time. For that matter, the *totum simul* all-at-once universe may not be impossible to imagine. It is a common view that it makes no sense to say that the entire universe could remain static for a certain period of time, on the grounds that if nothing changes, there would be no measure of how long that 'certain period' is. But Adolf Grünbaum[6] offers a thought-experiment. Since we would agree that a local *part* of the universe could 'freeze' for a while, and that all parts of the universe might freeze at different times, we can imagine that by coincidence, all parts of the universe might freeze at once for a while.

We do not have to go this far to hypothesize general simultaneity. Every formal or structural network functions simultaneously behind succession. Phenomenologically, we could not make sense of successive events unless we carried with us simultaneously a web of interpretative categories. And metaphysically, at least for Platonism[7], changeless principles are at the root of dynamic combinations.

(d) In addition to simultaneities in consciousness and in nature, others require cognitive or social construction. I mention five.

(i) Pairing: Husserl's fifth *Cartesian Meditation*[8] argues that a subject's experience of the unperceived backsides of a given object—sides that exist simultaneously with the front that she perceives directly—is correlated with the possible experiences of other experiencing subjects.

The simultaneity of perceived objects with unperceived objects depends on the simultaneity of the 'alter egos' that the perceiver is 'paired' with. Simultaneity requires that other egos be given not only transcendentally, but also concretely. Roads, monuments, clock towers, and meeting rooms testify to the presence of simultaneous intersubjective experience. Intersubjective simultaneity can feel like alienation, when communication devolves into simultaneous private thoughts (as in Ingeborg Bachmann's *Simultan*[9] stories), or it can feel like instantaneous communication across distance (as in André Breton's 'communicating vases'[10]).

One controversy concerns whether the simultaneity of perspectives involves exactly two face-to-face subjectivities (on Levinasian grounds that a third party interferes with intimacy), or else requires third person involvement (on Hegelian grounds that individual perspectives are socially embedded). Simultaneity can be seen as a single-termed all-togetherness, a two-termed relational at-the-same-timeness, or a three-termed intermediary being-with-others.

(ii) Dating: We often think of simultaneity as events sharing a date. Everyone of a certain generation remembers where she was on the day John F. Kennedy was assassinated. A calendar date has many effects.[11] It contracts a 24-hour time span into a single unit of reference in order to simultanize a set of events. The date thus constructs simultaneity not by matching an event and a temporal instant but by games of contraction (like Mark Twain's patent for a 'Simultaneity Game'[12]). It results in yearly anniversaries, which line up an original event with its celebration day. Dating may give the illusion of precision to events that are circumscribed only roughly. Since dates are measured by the rotation of planets, unlike the rhythm of most dated events, dating juxtaposes two very different time-lines. In contrast to dating mechanisms, adverbs like 'meanwhile' designate vague lateral simultaneities.

(iii) Scheduling: Marconi operators on ships like the Titanic were among the first to network global synchronization. Among other effects, this made global weather visible. The introduction of standard time in the late nineteenth century allows us to say that at the same moment it turns 6:00 in Guelph, it turns 12:00 in Paris.[13] Before this, there were of course correspondences between clock times in different places on the planet; but there were too many. Railway companies found it difficult to coordinate times in different cities, since each town would match its clock with the sun, and so it would have a unique time zone. Indeed, if clocks are set by the sun, Guelph time *should* be a minute or so earlier than Toronto time, since Guelph's position relative to the sun arrives simultaneous with a slightly different position relative to the sun in Toronto. The introduction of standard time, in order to simplify simultaneous settings of clocks, breaks the continuity of positions relative to the earth's rotation, by separating 24 artificial segments. We need to do this in order for efficient inter-city trade to take over from the urban flâneur who does not care what time it is, but it requires that international simultaneity operate not by instants but by blocks.

(iv) Plasticizing: Lessing argues in the *Laocoon*[14] that the literary arts (like poetry read aloud) employ temporally successive signs, whereas the plastic arts (like sculpture) employ simultaneous signs in space. Lessing prefers the art of succession, but by the era of formalism and structuralism, synchrony gets precedence over diachrony.[15] Lessing himself is concerned not with raw simultaneity and succession, but with hybrids. The problem for the Roman sculptor of the priest Laocoon was to capture simultaneously several stages of agony, to represent process in stasis.

(v) Stacking: When tones are put into a chord, they are simultaneous, and when the same tones make a melody, they are successive. Adorno argues that chord simultaneities in Schoenberg's music are more dynamic than those in tonal music (see Chapter 9).[16] The reason is that twelve-tone composition stipulates that all twelve tones must be played before a repetition of any note is allowed. This delays the return of each note as long as possible, to hinder expectation. But the series of twelve tones can be broken up into arbitrary segments, which can then be 'stacked' and played simultaneously. In traditional tonal composition, chord progressions are more or less predictable, and so simultaneity is in the service of succession; in music without a tonal centre, chords are unpredictable, and hence simultaneous through and through. One might think that unpredictable stacked simultaneities would reduce the energy of the music's flow. But Adorno hears the opposite effect: predictable succession does not make it feel like anything has happened, but the unpredictable leap from one stack to the next feels like a series of explosive simultaneities. Boulez and Stockhausen[17] serialize duration further, forcing us to listen to the simultaneity of different durations as much as to simultaneity of different pitches. With a similar insight, Apollinaire promotes poetry readings that he calls 'simultaneities', which become models for dadaist disruption politics and surrealist games of coincidence. In general, the early twentieth-century aesthetic of simultaneity breaks down continuity and energizes juxtaposition.

All of these constructions build a temporal substrate upon which time can pass, like a historical event that draws many causal lines into a singular happening, or a city plan with many levels of traffic circulation. What makes a phenomenon simultaneous is its ability to draw different temporal impulses into a singular temporal conglomerate.

Many problems are raised by these diverse phenomena: whether simultaneity is defined (1) by starting- and stopping-points or by indefinite duration, (2) by abstraction to a changeless time-line or by speeding up concrete changes to a level of unnoticeability, (3) by static or dynamic temporal maps, (4) by immediate natural identity or by artificial intersubjective construction, (5) by qualitative indifference or by quantitative combination, (6) by the conjunction and contraction of moments, the disjunction and repulsion of moments, or the equation and stabilization of moments, and (7) by the things themselves or by a frame of reference. My approach is to regard simultaneity as all of the above. The theoretical plurality counts as what simultaneity is.

Unresolved problems in the history of the philosophy of simultaneity: Husserl, Hegel, Bergson

Many sections of this book discuss simultaneity and delay in philosophers of the past. The particulars are not always commensurable, but there is a hermeneutical 'history of effects'. Concepts from one system are resources for others, particularly when systems break down and call for help. The truths in each combine into a multi-layered description of time itself.

I allow myself generalizations like the following:

Husserl grounds simultaneity in succession; Hegel grounds succession in simultaneity; Bergson grounds simultaneities in other simultaneities.

Husserl's account of simultaneity in his *Lectures on Internal Time-Consciousness*[18] (s. 38) begins with the structure of retention-protention, with its hinge in the transient now. The capacity to recognize that several objects *exist* 'all together at once' is correlated with our capacity to *experience* several objects 'at the same time'. The experience of temporal succession works by way of retentions and anticipations based on the ways that particular objects present themselves over time. Each object unfolds along its own temporal stream. So how are several objects experienced as being present at the same time? Simultaneity consists in the convergence of different time streams, and it is extremely difficult to determine what it means to say that a large number of complex objects exist as a whole simultaneously (see Chapter 2).

There is a staggered plurality in the now. Simultaneity is not an identity of what is present in a moment, but the differential between two or more time-sequences. We might have concluded that simultaneity can never be contained in a single now, except for the fact that the now itself is precisely this not-yet-coordinated plurality. Simultaneity operates in two ways. On the one hand, it 'stabilizes' flux into a congealed moment; a whole series of retentions and protentions is contracted into a singular moment of time. Simultaneity extracts a point in the history of each object from which to coordinate its other points. On the other hand, simultaneity coordinates situations in which one object has come to a natural stopping-point while another is in flux. Consider a pair of events with some shared and some unshared metrics, like a human walking beside a small dog in traffic: how exactly would we determine precisely what stage of the human's step is occurring simultaneous with a precise situation in the traffic jam?

Husserl's view is that 'simultaneity is never without temporal succession and temporal succession never without simultaneity'. But as we will see, both succession and simultaneity are undermined: succession because it is filled with delays, and simultaneity because it is filled with indivisible fluxes.

If pressed, this becomes a Derridian point[19] (see Chapter 7). The difficulty in lining up points in the temporal rhythms of a plurality of objects makes it hard to isolate the temporal location of any part of any object. The simultaneity of the parts of a single object will be subject to many-streamed rhythms, down to the smallest parts. It will not even be possible to name the part-object claimed to be simultaneous, since somewhere in that part are subsidiary parts with different rhythms. It may not even be possible to speak of retentions and protentions, since each state of a succession is a many-streamed staggering of fluxes and stopping-points. Husserl attempts to define simultaneity as a stopping-point in a succession of retentions and protentions, but succession depends on the very coordination of simultaneities that had depended on it.

The lesson from Husserl is that simultaneity is built out of the collection into unity of moments and rhythms in successive time-streams. The indeterminacy in the ordering of those streams is both the problem for simultaneity as well as the framework out of which simultaneity has to be rebuilt. But there is a different topos in the philosophy of simultaneity—one that does not draw simultaneity out of succession but succession out of simultaneity—namely, Hegel's dialectic (see Chapter 4).

Hegel's *Philosophy of Nature* (ss. 254–56) begins with the idea that extension reproduces each spatial point in a successor, and thereby traces points into lines, shifts lines into planes, and inflates planes into volumes. In Hegel's terms, each spatial determinacy is negated in its implied multiplicity, sublating abstract points in directionalities and coexistences. But while space has the structure of successive points along a direction, the multiplicity of places are given together and do not await future temporal movement. Indeed, if each point in space had to be seen at a separate time, or had to exist at a separate time, that is, if each point had its neighbours not at the same time but in a next time, there would be no space at all. Space is 'immobile collaterality' (s. 257).

The indifference of each point relative to each other is essential to the simultaneity of points. It is just because a point can continue in any direction, that all points on the plane exist at once.[20] This indifference of place to place implies the possibility of free movement. And freedom of movement in turn frees temporal succession from spatial determinacy. It is because space breaks up indifferently that points in space can be traversed over time. In this sense, the 'negativity' of space is time (s. 257). Simultaneity is spatial coexistence, but it is the time-form of space: space as the not-yet-altered coordinate, the alterable, already-altered-and-already-recoordinated. Manifold simultaneity is the kind of space within which something can happen in time. So (i) Time is the form by which certain lines across space are activated without activating all possible lines; (ii) Determinate movements are launched from singular points, so singularities in space are pregnant with temporal futures; (iii) Temporality moves across a multiplicity but is always located in singularity, and so always presents itself as an all-this-together-for-now.

Hegel shows how simultaneous points unfold into succession, just as Husserl shows how successive fulfilments contract into simultaneity. Idealism generally introduces lines of freedom into simultaneity—the freedom of geometrical possibility, the freedom of indifferent development, the freedom to move through a unified gravitational field, the freedom of matter to resist incursion, the freedom to mark and possess one's own region of space at the same time as others mark theirs. Simultaneity is a single-termed unity of all-togetherness; it is also a two-termed relation of at-the-same-timeness and a three-termed mediation of being-with-others. The problem that Hegel leaves behind is that if simultaneity provides the freedom in which all things are possible at once, it is a rigid freedom in which possibilities are all already laid out beside one another. We saw in Husserl that we cannot simply say that succession allows events to expand freely. Unless succession preserves the contemporaneity of events given together, it is too dispersed to count as an event. In a word, succession provides the context for simultaneous moments, but is too dispersed to give an account of time; simultaneity provides the framework for successive development, but is too tight. What is needed in order to loosen simultaneity so that it can persist through succession is not some characteristic of succession, but another level of simultaneity.

I use the vocabulary of tight and loose, contracted and relaxed rhythms, to suggest Bergson (see Chapter 6). For Bergson, time-consciousness is indivisible, but when we act, we see the results in space, and space is divisible. We slice up space, which makes us think we are also slicing up movement and time into instants. We then interpret each slice as a diagram of actions that happen simultaneously at that instant. Time itself is not divisible into simultaneous situations, but we read spatial divisions *back* as if temporal simultaneities were real.

In one sense, then, Bergson gives an anti-simultaneity account of time. But at least as important to Bergson[21] is the way in which memory coexists with present perception. Pure memory is not exactly simultaneous with any particular experience; it is quite the opposite kind of simultaneity—a permanent possibility simultaneous with every moment of psychic life from then on. Whereas some kinds of memory trace backwards from the present, one step at a time, Bergson is more interested in the kind that picks out any event we like from any time period without regard for order. This is 'pure memory', the totality of which is always available. Bergson concludes that memory itself is never lost, though it is a 'virtual' feature of consciousness rather than an actual one. Bergson's twist on Einstein is that simultaneity is the contemporaneity of different frames of reference calculating simultaneity differently: different tempos in the same action, some divisible and some indivisible, some multiple and some singular, some passing and some permanent, some real and some virtual. Bergson does not have a general theory of multiplicity. It is Deleuze who casts Bergson's theory of time in these terms.

Deleuze in *Difference and Repetition*[22] asks what makes it possible for the present to pass. A present moment cannot pass out of existence simply because another moment arises. But there is nothing in a present moment that would lead it to pass—on its own, it would remain present. Deleuze's solution is that each moment has two distinct temporal statuses—it is present and past—at the same time. On the one hand, an event takes place (which defines the present); on the other hand, it remains available to have effects afterwards (which defines the past). Ongoing, it is present; but qua given, it is already past at the same time. A moment does not pass out of the present in order to become past; its pastness is contemporaneous with its being present.

In short, present and past are not two stages of temporal development, but are two distinct structures of time. Since an event's past is contemporaneous with its present, *all* of the past is contemporaneous with each new present. Events still appear successively, but each in different ways 'virtually' plays out the life of all pasts (83). The present plays out yesterday's events and last year's events simultaneously. The present enacts the past as its own distant contemporaneity. Hence, an event in history, like Joan of Arc's military-sexual-theological rebellion, has running effects centuries later.

The history of the concept of simultaneity provides many trajectories for thought that we will follow through. There are clearly many structures of simultaneity: a simultaneity of the present that anticipates and retains; a simultaneity of the past that stores coexisting events; and a simultaneity of the future as a permanent possibility for event interaction. Some simultaneities arise from succession; some are indifferent to it; some generate it. As each philosophy of simultaneity reveals difficulties, we cycle through the others to find solutions. But as the problems mount, we also need theories of failed simultaneity: namely, theories of delay.

We need a theory of the many levels of simultaneity and delay to explicate how there can be two events in one time, how there can be one event spread over many times, and how all conceivable variations on events and times can play out.

(b) Delay

If simultaneity in all these ways contains different times, bringing events, rhythms, and projections at different times into one time zone, then the simultaneity that exists at one moment is not completed at that moment. Simultaneity is by nature delayed, and delay in turn is differentiation within simultaneity.

What does it mean for an event to be delayed? When a flight is delayed, its actual departure is compared with the moment when it should have departed. It was supposed to depart simultaneous with certain other events, and simultaneous with a reading on the clock, but did not. In some cases, as in delayed reaction or delayed gratification, delay is part of the normal flow of events; without delays, all the phases of an event would happen at a single absurd moment. The flight's delay has meaning only as the stretching out of an expected, virtual simultaneity, just as simultaneity has meaning only as the synchronization of delayed reactions. In other cases, though, delay ruins the coherence of an event, as when the meaning of a communication is endlessly deferred, or just rewards are indefinitely delayed. Delay sometimes refers to events that never arrive. But it sometimes refers to events that cannot stop arising, or to ends that cannot finish ending. Sartre describes the end of the occupation of Paris as 'the end of this war that cannot get its end to end'[23]; Baudrillard defines the end of history as history that 'cannot even manage to come to an end'[24].

Some delays are finite and pragmatic; others are infinite and transcendent. Some say, 'wait'; others say, 'whenever'. Some are subjective, in relation to expectation or boredom. In some cases, whether there is a delay or not depends on perspective: what is delay for a patient may appear on time to the doctor. What reasonable people call 'on time' differs across cultures.

Other delays are objective, as in the metaphysics of gradualism, the semantics of hypothetical conditionals, the physics of causality at a distance and the problem of the time arrow, the technology of time-delay medications, or the practices of catching up. There are many closely related concepts of delay, deferral, deference, postponement, hold-up, and slow-down, and there are delay phenomena in ontology, theology, aesthetics, phenomenology, cognitive science, medicine, psychoanalysis, and politics. Different languages have different classifications. French distinguishes waiting periods, intervals, and the time it takes to get something done (*dans un délai*) from excessive slowness, or being behind the times (*retard, attardé*). The German *Nachträglichkeit* suggests time lag: an event dragging behind. Just as simultaneity can refer either to two simultaneous events or to a single event in its own proper time, so delay can mean either one event set back by another or a single event not in its proper time.

Like simultaneity, delay has a spatial component as well as a temporal one. Zeno's paradox of Achilles and the tortoise does not hang on how fast or slow the runners go, but on the number of spaces that Achilles has to pass through. At this level, the paradox is about space more than time. The striations of space create delay. But if there were no time, Achilles could be in all spaces simultaneously, and would already have caught up. At this level, the paradox is about time. Achilles can never catch up, or cancel the delay. There will always be a remainder still delayed, an infinitesimal division, a history of failure, a mark of hasty recovery, a strategic difference. Achilles can get to the end as fast as he wants, but he will never catch up with the other racer. Delay is the other in time.

Whether the delay is suitable or ruinous, it is difficult to define delay in purely temporal terms. Since events occur in order, there is a sense in which every event occurs at the appropriate time. What is the status of delay? Simultaneity is a convergence of time-lines, a mark of the coexisting rhythms of related events; this property of simultaneity is the source of the ontology of delay. Delay occurs when a single event both extends away from its starting point, yet is prevented from progressing to its conclusion. In delayed reaction to a traumatic event, for example, the trauma gets stuck in time, simultaneous with every present, perpetuated all along the time arrow. Memory in general is a delayed reaction: giving meaning to succession, tying diverging events to a common starting-point, and prolonging events into an indefinite future. Delay is the extenuation of temporal flow, but also a condition for temporal continuity. Delay represents the broken-up singularity of a simultaneous moment, but its staggered rhythms locate a complex together in time.

Delay does not necessarily refer to the future, since delays have occurred and have been resolved in the past, but the unsatisfying character of delay does relate some moment to *its* future. The anxiety of delay holds the future hostage to the present momentum originating in the past. Delay installs the future in the past, brings the past into the present, and endows each future with more futures. If we become sceptical of the existence of the specious present, past, or future, then delay will have no referent. Or to put it conversely, delay ties past, present, and future into a single web in a way befitting the speciousness of their independence.

Again, we begin by gathering phenomena. In ethics, delay arises in unfulfilled promises and inefficient political systems. In medicine, pharmaceuticals have time-delayed effects. In psychology, consciousness is slow to pick up input from the external environment. In cubist aesthetics, the artist prevents synthesis from building up. In theology, delay is rendered visible by prophesy. In practical affairs, bureaucracy means delay. In phenomenology, there are unfulfilled hopes, frustration, and procrastination, when we do not know if something is happening, is going to happen, or is supposed to happen. Pragmatically, some delays are good, like a reprieve for the condemned prisoner or nuclear deterrence. In political economics, there are interest charges payable later, the futures market, and the expectation of profit from investment. Capitalism may eventually pay its dues with a communist revolution. The unfolding logic of dialectic is an economy of delay.[25]

Sometimes a delayed event has not happened yet, but ought to. The 'ought' plays a tricky role in the definition. Something is delayed not just when two things happen at different times, or when one has happened already and the other has not happened yet. The event is delayed if it should be happening now, either in a moral sense or in relation to expectation. 'Delay of game' is penalized in hockey. Being lulled into a false sense of security is dangerous. Something good can happen too late, due to procrastinators and laggards, lack of preparation or over-ripening. Alternatively, something good may arrive too early, like a marriage proposal; most bad things arrive too early. Sometimes delays ought to be avoided, but not always. Delaying medical diagnosis is a bad idea, but delaying treatment of prostate cancer is recommended. 'Deference' to authority can add time for careful decision, but not if it leads to paralyzing deferral.[26] Sometimes the presumed norms around delay are illegitimate, as in the old attitudes towards 'retardation', or 'backwardness'. Some delays have positive value: 'slow food', fashionable lateness, civil disobedience slowdowns, minimalist music, or when something bad is held at bay. Fortunately, the Doomsday clock has stuck for years at a few minutes before twelve without arriving. Sometimes, delay has the positive value of rupturing the always-already. Though it issues advance notice, delay's disruptive quality distinguishes it from normal teleological schedules.

There may be purely physical delays, as when a motor stutters. Urban designers find that 'in rush hours a double-parked car blocking one lane of a main road for 12 minutes can delay 2,800 other vehicles'.[27] In time-release medications, or in delaying gratification over immediate pleasure, delayed effects do not frustrate expectation but are exactly what is expected. Delay does not always mean disappointing the expectations of an observer; it can mean that a thing is falling behind its natural pace, alienated from its nature, whether it is capable of returning to its own speed or not.

There are delayed *reactions* to provocative events. For Freud, things can happen to a child who does not know enough to react until years later. Delay is the mechanism where something is rendered virtual and stored in the unconscious for later.

Delays have different forms of resolution. For some, resolution is prepared for and definite. If I cannot meet a deadline today, it will be done tomorrow. Others are indefinite, halting, or up in the air. Still others are infinite, since the sphere in which they could get resolved is different from the sphere in which events happen. On some readings of the Last Judgment, for example, the end of the world and the restoration of all to God, is not a chronological end to the physical world, but an indication of a transcendent realm. In still other cases, delays exist because the time measure is out of line with the material it measures. For example, Morton Feldman's minimalist music is ultra slow. If played at a normal speed for classical music, it would seem as though several notes were played at once, but played as intended, far slower than normal, we hear all the variations of tempo. There may be one measure of time for the first musical event, then a slower measure for the follow-ups, in addition to the everyday pace at which the listener lives. If classical music is played either too fast or too slow, the natural delay in the resolution of dissonant chords is lost. If the resolution chord arrives too soon, we hear no tension; if it arrives too slowly, the dissonant chord does not seem resolved at all. In a different way, meditation slows things down so that events will not feel like they call for follow-ups. The goal is to cease feeling dissatisfied by the fact that we have not yet achieved happiness. Creating multiple steps out of a single act, working-to-rule, can be a mode of civil as well as conceptual disobedience.

There is a myriad of different delay structures. There are legal liabilities for costs incurred by construction delays, injustices arising from delays in the courts, rain delays in baseball. Mathematics deals with 'delay-differential equations'[28]: 'the derivative of the unknown function at a certain time is given in terms of the values of the function at previous times'. The International Air Transport Association maintains 'Delay codes' to track check-in delays, standby policy, and arrival times. Many books with titles like *Is God Keeping You Waiting?* are sold on Amazon. Sociologists study queues and wait times. What used to be called 'tape delay' shapes communications, computer interface, and the recording and playback of images. Digital delays predetermine operations to be carried out at set times, controlled by 'gates'. Mechanics depends on latent effects. Administration manages workflow. Drag racers in the 1990s cheated with 'delay boxes' that would keep the car stationary just before the start of the race with the engine at full power. Childhood social development requires that (human and other) children learn how to delay gratification. Plato thinks that the fact that gratification is delayed implies something wrong with pleasure. Don Juan's pleasure is deferred just because it is never delayed. Peter Pan won't grow up. For Romanticism, it is human to suffer from deferral, with longing and patience.

Sometimes delay covers the whole event, which can all be put off together. Sometimes one part of an event takes place while the other does not; a bomb on a fuse is calibrated for delay. A screenplay is optioned, but the filming delayed. A blueprint exists, but not the building. The important thing is that the period of delay is not a temporary situation whose resolution will bring a lasting situation; delay is the real condition of situations throughout their temporal duration.

What does it mean to catch up after a delay? If two people are running in step, and one falls out of step and falls behind, how should we describe exactly how she falls back into synchronization with the other? It is easy to see how she accelerates to catch up, but how exactly does she fall back *into step* and not accelerate past the other? If one falls behind in one's education, one can read more but may not ever precisely catch up to where one would have been if one had not fallen behind. Once delay enters the picture, the picture from then on exhibits that delay. It is difficult to say whether re-synchronization is the normal result or a rare one, whether Kierkegaard is right to wait for Regina to be returned to him. 'Making up for lost time' (in French, *combler son retard*) tries to fill the hole of delay, and in some cases, it might even surpass expectations. Delay is not just a time-debt to be repaid; it gives the potential to remake oneself, to surpass the norm. Delay is not always *temps perdu* but can be time saved up.

Our second approach is to analyze the concept of delay through the history of philosophy. Often it is inside a treatment of simultaneity that we find a philosopher discussing delay. But there are also implicit concepts of delay in the ancient relation between time and eternity, in the late modern introduction of mobility into reason, and then explicit concepts in late twentieth-century motifs of deferred meaning and incompletable politics.

For Husserl, all experience depends on anticipation, and this may be the deepest argument for the universal structure of delay. All sides of an object like a cup coexist; as objective 'noemata', they exist simultaneously. But my experiences of the cup are successive; the subjective 'noeses' occur one after the other. For the subjective and the objective to line up, it has to be possible for the successionist description, the bit-by-bit, one at a time, temporally spread out description, to add up to the objectivist, all at once, description. From the subjective perspective, more is anticipated than is actually present at a given time. Verification requires that *most* of an experience be *not actually* experienced but only virtually, or potentially, or futurally, anticipated. To give meaning to one experience, we have to imagine an infinite number of further experiences. But it is not exactly a question of images. Anticipations are not all explicitly conscious, though they are constrained by experiences that are conscious. The other side of the cup is not going to be ten times bigger than this side. Similarly, I have a fair range of expectations of what you, dear reader, are likely to say about this chapter. You could always surprise me, but the possibilities of unknown futures, even ruptures, are also built in as anticipations. In short, most of experience consists of expectation in the form of not-yet.

Setting up anticipations is correlated with fulfilling or satisfying them. This could be called temporal mood. The other sides of the cup I see are correlated with my moving to the next side to see what was expected. In everyday experience, we are constantly confirming (or falsifying) our expectations. So the fact that most of an experience is off in the future does not undermine the epistemic solidity of the world. On the one hand, the whole cup is never going to become present all at once; there is always delay in its presence. But the fact that we anticipate running through the sequence of perspectives, all of them and again and again, means that there is an ideal endpoint which we always have again, anticipatorily in mind. So at any point in the process that we say, 'I better check the back again', we can go back to the backside. Delay is a crucial condition for verifying that the world is real.

For Derrida, the problem is that the ideal endpoint does not remain constant. Sometimes in the course of small confirmations, we not only realize 'I expected this but got that', but also 'the expectations that led here were not what I thought they were'. Worse, if we find unanticipated mold at the bottom of the cup, we wonder if we have been subliminally anticipating mold all along, and wonder 'why have I been choosing to ingest this disgusting mold?' It is not just that we are going to change our expectations now; it is that all along, we had strange expectations, which we did not realize we had. Husserl would like the ideal endpoint to synthesize a continuous series of experiences, even if the endpoint is always deferred. Derrida's claim is that if the ideal endpoint is never present, then the stream of experiences that supposedly leads up to it is not continuous but fragmentary, loose-ended, or even contradictory. It is not merely that we cannot articulate in advance all that we expect of a given object. It is that we do not know where any experience is leading, or what any memory meant. Any stage of any interpretation of anything along the way to anything has a delay character that produces uncontrollable possible futures. In short, delays cannot be resolved.

Derrida upsets traditional categories of past, present, and future; the upside is that this makes the past open to future revision. It is ontologically inevitable that the past will be not just reinterpreted but reproduced in a new way. It follows that the future is not entirely future, that the future is partly a construction of the past, and that while the future never arrives, it has always already been present in that very state of deferral.

The history of delay theory gradually gathers every ekstasis and level of time under its categories. Both simultaneity and delay (no less than their common enemy, succession) cover the full range of temporal phenomena. It is not that some phenomena fall under one of the two categories and others under its rival: all phenomena are eventually gathered under both. The co-extension of simultaneity and delay constitutes the nature of time.

Our third approach will be to deduce consequences from principles. A delay is an event that to some extent has happened and to some extent has not. It is an event with two times in it, one event split into two.

We can summarize the problem of simultaneity and delay in this way: When many things are happening at once, different parts of the various events take place on different time scales, and this conceals difficult problems of coordination. Consider trying to coordinate the various timescales when a person walks down a street chewing a sandwich, while a bell goes off and someone throws water off a balcony, which hits the ground a moment later, at exactly the time the person is at a certain point in taking a step, chewing the sandwich, and hearing the second verse of a song on iTunes. There is a simultaneous moment at which it all happens at once. But at-onceness involves different processes with their own rhythms coinciding. Simultaneity is constituted by the crossover of differential time flows, each with its own anticipation and retention patterns, some densely packed with short sub-events, some drawn out, some temporarily static. It is not clear that a theory for coordinating time-flows is even possible, or which philosopher offers the best resources. The theory would have to describe those inevitable unsynchronized sub-sequences across broadly simultaneous sequences: staggered time-lines, differential time within the schema of 'at the same time', delayed simultaneity. Simultaneity and delay turn out to be not so much contraries as alternating descriptions of the same complex process of temporal coordination and re-ordination, of synthesizing and diverging time-lines. Both simultaneity and delay are necessary conditions for the possibility of time. There would be no time, no succession, if all changes occurred at exactly the same rhythm, at the same periodic intervals, with no intervening disturbances, no partial completions still to be resolved, no building up or gearing down (hence no beginnings or endings), no micro-divergences into smaller segments, no alternative futures marked in the moment itself. If every event occurred in the same intervals on the same rhythm, one world-situation would last for a certain amount of time, 'at' a time, and be over; then the next would be in place, all through the universe for the same amount of time; and then the next situation for the next amount of time, as if a movie were going by in equal frames, as if Zeno's arrow after all did not move. Would that be time? It is not clear how a perceiver would see the continuous transition from one to the next, how an event at one time could transmit its causal force onto something in the next, how any event could last for more than an instant, how different events could last different amounts of time, or how a particular event could be individuated from a background. Without the possibility of interposing connectors between two beats, it is not clear what it would even mean to call an event 'next'. Time requires overlap as much as division, and overlap means a plurality of rhythms. It means that one situation keeps going *while* another starts up and that succession is a dialectic of simultaneity and delay. In short, an event needs alternative time-lines coexisting with it: delays that are simultaneous with it.

There is not just delay of the next, but delay within the now.[29]

Chapter 2

Phenomenology of Simultaneity and Delay: The Problem of Converging Timelines

Theme of the chapter: Simultaneity is constructed as a convergence point across multiple anticipation-fulfilment time-lines. The completion of such convergences is delayed.

I will generate two problems of time in Husserl. The problem of simultaneity is how to construct simultaneous intentionality out of convergent time-flows. The problem of delay is how, after an experience has been fulfilled, we can remember how it looked before it had been fulfilled, or in other words, how anticipations are retained, or how the replacement of the past by the present can be successfully delayed. In Chapter 7, I discuss Derrida's solutions.

(a) Simultaneity in Husserl

How are two or more objects or events experienced simultaneously? The very existence of simultaneity in experience is not self-evident. One might argue that simultaneity is reducible to switching attention between objects in rapid succession. However, phenomenological description should, as a first resort, try to save simultaneity as a genuine phenomenon, even if, in Husserl's words[1], 'simultaneity and temporal succession must become constituted correlatively and inseparably' (s. 38, 78).

Husserl analyzes simultaneity in four pages of the *Lectures on Internal Time-Consciousness*, s. 38: 'The Unity of the Flow of Consciousness and the Constitution of Simultaneity and Succession'. For Husserl in general, the time of the flow of consciousness should be synthesized with the time attributed to objects. Recognizing that several *objects* exist 'all together at once' should be correlated with having several *experiences* 'at the same time' (77). Our sense of inner time consists in our ability to retain experiences of objects seen in the past, hold them in present consciousness, and anticipate future continuations. The way we identify objects is that each experience of an object refers back to memories of it and forward to further possible experiences of it. This structure of retention-protention[2], with its hinge in the now, is in one sense a formally permanent structure of consciousness and is in another sense tied to the matter of each particular objective experience.

There is a tension in Husserl's Time-Consciousness lectures. The general approach of this text is to emphasize the subjective flow of time and to describe the experience of nows flowing past whatever their content. But most of Husserl's texts emphasize objective intentionality and describe the way we keep track of the history of our perceptions of objects and our perceptions of the histories of those objects. The former approach describes subjective time formally without concern for which objects we experience; the latter ties our experience of objective time concretely to the particular objects we experience. If we could define simultaneity in formal terms alone, as anything that occurs in the same now, there might not be any problem. But a problem arises because the subjective time-flow is never entirely independent of intentional references to the time-patterns found in particular objects.

Even subjective retentions and anticipations are based on individual objects presenting themselves over time. My sense of objective time is built out of seeing a bird fly by, watching hockey on TV, or waiting for onions to caramelize. My experience of succession in each case is an experience of the time-line of a particular object. Because of this, Husserl acknowledges that the temporality of consciousness 'breaks down into many streams'. How is it, then, that several objects can be experienced as being present at the same time? How can many objective lines of succession be present in a single subjective simultaneity? If we could posit abstract slices of simultaneous time on a time-line, we could avoid the problem. But for phenomenology, abstract time is derived from our intentional experience of objects in the world around us as they persist over time. The problem of multiple rhythms in the anticipation and fulfilment time-lines for different objects is a more serious problem for simultaneity than Husserl thinks.

'Same-now-ness'

I may understand two objects in my experience to exist at the same time, even though I do not experience them at the same time. To take Kant's example, when I walk around a house, I experience each room at a different time, but my experience is that they all exist at the same time. Conversely, there can be two events that happen one after the other, which I learn about at the same time. Husserl thinks it may be hard to establish which objects exist at the same time but easy to see subjectively when we are having two inner experiences at the same time. I will argue that difficulties on the object side complicate the subject side[3].

I emphasize Husserl's schema of anticipation and fulfilment, though he does not use it in *Time-Consciousness* as much as in other texts. The reason I can have a continuous experience of this cup is that when I see a new side of it, it is more or less the side I expected to see when I turned it. The new side fits into a pre-existing scheme; it fulfils an anticipation. Anticipations are tied to particular objects. The general expectation 'something is going to happen' is not specific enough to generate continuity. Similarly, retention of the past cannot take the overly general form: 'I know something happened in the past, but not what'. To describe what is in the now, I locate where I am relative to objective retentions and protentions[4].

So how do I experience two objects in the same now? Suppose I want to know what stage I have reached in bringing a coffee cup to my lip at exactly the moment that I have reached a certain point in reading a page of a book. The point I have reached in carrying out anticipation-fulfilment chains relative to the cup has to be matched up with the point I have reached in carrying out anticipation-fulfilment chains relative to the book, and so on for all the objects I am simultaneously aware of. This is evidently quite a task. The question whether simultaneity is possible hangs on whether the several chains have commensurable points that can be mapped onto one another. In experiencing the cup, for example, my expectations have to do with how long it takes to turn the cup around or how far I can turn it before the coffee spills. For the book, I anticipate how long it takes to flip through pages, how long it takes to read a page of Husserl, and so on. The two sets of expectations contain different measures of how the future is constructed. And that is just for two objects close by. Walking alongside buildings, gazing at the environment, eating sandwiches, singing to my child, each with its own complex rhythms of anticipations and fulfilments, all have to line up if I am to experience one slice from each of those processes at the same time. Describing what the experience of simultaneity is like, and how it is constituted, is difficult just because the joint anticipation is built up one object at a time.

The history of anticipations and fulfilments is crucial to locating each now, that is, to determining whether a given content exists in the present or in some other time. It is not quite that without a history of anticipations I would not know that it is now *now,* but without it, I could not pick out together all the things that are in the now.

Husserl does not think this is a devastating problem. He figures that whatever the objects are around me, I experience them all at the same time: 'no matter how many objectivities are constituted in the now, they all have the same temporal presence' (s. 33). Indeed, he says, as time flows by, all contents recede into the past at the same rate. 'They retain their simultaneity in flowing off'. All the stuff around me right now seems now; in one second, it is all going to seem one second old.

This is why he starts s. 38 with the principle:

> Now, in reflection we find a single flow that breaks down into many flows, but this multitude nevertheless has a kind of unity that permits and requires us to speak of *one* flow. We find many flows because many series [*Reihe*] of primal sensations begin and end [*anfangen und enden*] (76).

This introduces the next crucial point. The reason we experience many flows is that we experience time serially, with breaks, that is, with beginnings and ends. We experience the flow of sitting in the lecture and then the flow of going out for coffee. In one sense, it is all one flow, but in another sense, one flow moves along and takes a natural break, after which the next flow takes over. There are overlaps among flows, experiential subroutines, but there are natural breaks at certain points in the flow.

The many-ness of the flow again pertains to the different objects of experience. Yet Husserl's recurring move is to cancel the plurality of intentional objects in favour of immanent unity. To pick up where we left off:

> We find many flows because many series of primal sensations begin and end. But we find a connecting form because the law of the transformation of the now into the no-longer—and, in the other direction, of the not-yet into the now—applies to each of them, but not merely to each of them taken separately; there rather exists something like a common form of the now, a universal and perfect likeness in the mode of flowing (81).

Husserl finds the unity across flows in the form of the Now. Of course, Husserl is not speaking of the instantaneous now that many philosophers say vanishes but about the now as the process moving from not-yet to no-longer. But how does process guarantee that there is just one flow? There may be a single law mandating that things pass away, but why does one law imply one stream?

It is in this context that Husserl introduces 'simultaneity'. In the common flow of the now, 'many primal sensations occur 'at once' [*auf einmal*].

When any one of them elapses, the multitude elapses 'simultaneously' [*zugleich*, 77; elsewhere *Gleichzeit*]'. All changes in experience feel like the same kind of change. But this is the first time Husserl says that many sensations occur at once. Does this follow? Even if all things that pass by do so in the same sense, must that mean that at any given moment several things are passing by?

If there were no simultaneity in the now, Husserl would not be able to argue that experience is unified into a single temporal stream. The act that formally gathers all our experience into simultaneity in the Now is also the source of unity in succession. But if simultaneity guarantees continuous succession, the succession of beginnings and ends makes simultaneity possible in the first place. Pinpointing a moment of simultaneity across two processes requires that in some sense, the two processes have a common interim starting- or stopping-point.

'Stopping-points'

Husserl's conclusion so far is that all successions operating during the Now exist 'at once' (*auf einmal*). 'Same-time-ness' means 'same-now-ness' (*Gleich-Jetzigkeit*).

Simultaneity is, first, everything that exists together in the Now; and second, it is what elapses into the past together. The notion of elapsing, of a flow coming to a natural stopping-point, is crucial to the 'at once'.

> We can extract a point in the continuity of modes of running-off [*Modis der Abgelaufenheit*: literally, 'running off'; figuratively, 'elapsing', as though a stream of time ends by running down the drain; also *aufhören*], and we then find in this point too a being-together of running-off modes perfectly alike in form; or rather, we find an identical running-off mode… [This] is the site for the constitution of simultaneity (78).

If there were ongoing flows, but without elapsing, the flows would occur *throughout* the same period, but there would be no 'at once'. For the latter, we must be able to stop at some point, and say: now it has come to the point when two processes coincide.

But again, this is problematic because different objects have different times at which stages in their unfolding become noticeable. In some cases, there are natural stopping-points at which the flux of an object's mode of elapsing will be noticed. Fiddling with a cup involves continuous expectations, whereas reading a book gives expectations in quantum chunks, since we read discrete pages and not the spaces in-between. We do not normally look at pages in mid-turn; there is nothing there to look at. If two objects have the same kinds of stopping-points, it will be easy to say, when they get to those points, that two events happen simultaneously. In a chord progression where each tone in each chord gets one beat, the voices share starting points, so all the tones in each chord are noticeably simultaneous. But if one process stops while the other is in the midst of flux, it will be difficult to say what in the latter's continuous movement is simultaneous with the former discrete event. Each object (indeed, each part of each object) in each context has its own gradations of temporal stages. To be sure, all objects pass before an experiencer and pass *through* time at the same time. In Bergson's terms, they are 'contemporaneous'. But simultaneity requires more, namely that the definable phases of the two movements coincide.

'Tempo'

Husserl calls this the issue of 'tempo' (77). (Bergson calls it 'rhythm'.)

How is it possible to walk and chew gum at the same time? Sometimes, when I am walking by a building eating a sandwich, my chew cycle falls into sympathetic vibrations with my step cycle. But what should we say about simultaneity when they are not in sync, and when meanwhile, somebody throws something out a third floor window which I only notice when I hear the splat just barely missing my head? I do not know exactly how fast each experience is changing. My chewing has a rhythm, my singing has a groove, and my thinking is in a rut. I have vague expectations about odds and ends, some in clusters. Do the taste at mid-chew, and my semi-conscious sense of unidentified matter whizzing by, really elapse together with the stopping-point of my last step? Without that synchronization, there will be generalized flow, but no 'at once'.

> The 'being together' [*Zusammen*], the 'being-all-at-once' [*Zugleich*] of actually present primal sensations is all-embracing... (77).

In s. 31, Husserl says that we have to experience objects in their context [*Zusammenhang*], as an environment, and this means that we have to experience objects all at once as one chunk. The whole environment elapses, 'in absolutely the same mode with absolutely the same gradations and with absolutely the same tempo'. Here is the problem: whether things do indeed recede into the past with the same tempo.

To continue Husserl's sentence, objects in a given now pass in the same tempo,

> *except* that, in general, one ceases while another still has its not-yet before it—that is to say... *except* that the series of primal sensations... are variously prolonged, corresponding to the varying durations of the immanent objects. They do not make use of the formal possibilities in the same way (77).

I think it is possible to prove transcendentally that there must exist staggered processes, since if every event marched to the same stopping-points, there would be no way to tie over the beats, and the continuity of time would be lost. Husserl acknowledges that different objects are 'prolonged' in different gradations, but in spite of that, he thinks there is a single tempo for all objects. At several points, Husserl comes close to acknowledging different tempos. (i) In s. 31, he speaks of a 'joint positing' of 'environmental intentions'. (ii) In s. 41, he says objects often present themselves punctually in quantum changes (in *Zeitpunkte* and *Zeitstücke*), and that there is an 'ever-growing divergence' across the different streams. (iii) In s. 35, he says objects appear according to 'alterations of speed or acceleration (as it were)'. (iv) In s. 29, he says that it is a challenge to 'find the thing' in the 'possible trains of memory'. By contrast, in Appendix VII, he says we experience time 'one dimensionally' at 'the same speed'.

Without stopping-points, nothing would cease *while* something else is still in the flux of its not yet. An opera ends while I am thinking about supper. The music ends at a certain time, but my thinking about supper has not reached any stopping-point. It has no specific time to be pinned to. There is no deep problem as long as one event finishes while the other has a measurable amount of time remaining. And no doubt there are many delays, rubatos, reverbs, stutters, and so on, which can be resolved by technical means. But since every object undergoes some processes in-between its natural stopping-points, strictly speaking there are at any given time, measured *as* a given time by the tempos of *certain* objects, *other* objects that simply do not have a measurably given time at that time. If we ask precisely what state the hunger was in when the aria began, there is, for transcendental reasons, no determinate answer. And if we ask whether a subject experiences the whole environment simultaneously, that too is undecidable. There is nothing indeterminate about objects, subjects, or processes, as defined by their own appropriate rhythms—the only indeterminacy concerns the judgement of simultaneity.

Simultaneity, in short, is not an identity of what is present in a moment, but the experience of a staggered differential between two or more time-sequences. Phenomenology cannot discount the fact that we experience simultaneity. Simultaneity's strangeness belongs to the things themselves. The now is the pre-established harmony of the unharmo-nizable.

Simultaneity has two functions. Sometimes it does extract an overall stopping-point. Other times, it coordinates different sorts of stopping-points. These two functions generate a distinction in Husserl's terminology. *Gleichzeitig* is his word for relative simultaneity, roughly coordinated flows and stopping-points. *Simultan* (Appendix V) is his word for the limit case of perfect coordination, as when one hears a tone C played on one instrument at exactly the same time as one hears it played on another instrument[5]. These two C's have such identical retention-protention time-lines that we hear the two events as one and the same note. It is necessary to go to this extreme to find perfect simultaneity in a single tempo. Indeed, the more perfect the convergence, the less there is a question of *two* processes being simultaneous.[6] *Gleichzeitigkeit* is only called simultaneous because it approximates *Simultan*, but *Simultan* reduces to identity unless it partly breaks down into *Gleichzeitigkeit.*

This calls for rich analyses of 'rhythm' (see Chapter 9). We tend to think of rhythms in binary groupings like 2/4 or 3/8 time, where two rhythms are easily commensurable. Yet 'new' classical music adopts 'irrational' rhythms, changing tempo virtually note by note, calling for one instrument to play a few notes in 13/21 time while another plays in 12/27 time, and then for each to switch to another tempo in staggered ways, occasionally by chance, or too fast for the performer to control, to the point where the listener cannot be certain whether tempos coincide or not.

Wundt's *Outlines of Psychology*[7] (an influence on Husserl) describes 'isochronic' situations, as when, in walking, a person's arms and legs are in movement at the same time. Coordinating arms and legs is less a case of perfect simultaneity than of 'rhythmical movement'. At the beginning and end of each step, there is a 'feeling of *fulfilled expectation*', and during the step, '*strained expectation*'. In the repetitions, 'the beginning of the first of two successive periods is emphasized' (the first step more than subsequent steps). Only later might the 'movement begin to be metrical', if stepping becomes marching or dancing. Simultaneity does not just consist in the fact that the beginning and end of two events are identical. It involves periodic strain, unequal valuation of sub-events, and interplay of shared and unshared metrics.

At a certain point in the description of simultaneity, Husserl says, one can only say, 'Look' (79). Husserl thinks introspection immediately exhibits primal simultaneity, but I think the injunction to 'look' has a second phenomenological force. The question whether two things are simultaneous has no simply correct solution, but a range of experiments that can be performed from different frames of reference.

'Before-simultaneity'

The injunction to 'look' presupposes that the grounds for simultanization are pre-given. 'The 'being-together', the 'being-all-at-once' of actually present primal sensations is all-embracing: all-embracing too is the 'before', the 'having-gone-before' of all the immediately preceding primal sensations' (77). For Husserl, remembering that several objects were simultaneous depends on the simultaneity that existed in the original now. Husserl calls this the *Vor-Zugleich*, the before-simultaneity, the before all at once. If we did not have in advance the expectation that books have pages that exist all at once, then there would be no tempo for scanning a page. Prior simultaneities are tempos set up in advance to lead to an at once.

In this way, experiencing simultaneity takes a 'step back' (*Zurücktritt*, 79)—both in that the experiencer withdraws attention from single rhythms to see several objects at once, and also in that the simultaneity of objects depends on their common past. Simultaneity it not something we could construct from scratch out of anticipation-fulfilment chains without getting lost in their overwhelming complexity. We can only experience simultaneity if it is posited retroactively, if we can posit a priori that all world processes overlap in spite of not lining up. This opens up, as a region of phenomenological research, the layers of staggered time-lines.

Husserl concludes that, 'simultaneity is never without temporal succession and temporal succession never without simultaneity' (78). In section 39, he compromises with multi-tempo simultaneity. After noting that objects move by at different speeds, Husserl says that immanent flow constitutes 'transverse intentionality'. We can always create a crossover intention that simultanizes across tempo-lines. I call this 'delayed simultaneity': simultaneity dispersed into succession but held elastically back in unity. It is not that consciousness's hunt for simultaneity fails. It obviously succeeds. But it succeeds just because when processes do not line up at common stopping-points, we can reassign different moments of different processes so they slide into place. Page-turnings are discrete, and cup-fiddlings are continuous. But we can nevertheless recall the discrete first moment of picking up the cup, count a discrete number of pages read since then, then approximate a quantity of semi-conscious fiddling in the meantime, and so by pointing back to a common starting point earlier on, and looking forward to a time when we will finally put the cup down, we can overlay one discrete event capable of precise simultaneity with one continuous event not capable of precise simultaneity, and do so in a single temporal intentional reference, that is, simultaneously. Strictly speaking, once the cup starts rotating, it is already too late, and still too early, to line it up with a single turn of a page. Until the next discrete event in one process, its capacity for simultaneity with the other process is delayed. But because of the way memory and anticipation are built into the presence of processes in the first place, these delays do not block the apprehension of simultaneity, but rather produce it. We make simultaneities visible and memorable by shunting intentional objects around, stretching them into continuities and slicing them apart, holding them back and letting them run off, rewinding their clocks and waiting for alarms to go off.

Many cases contribute to such an analysis. When someone says something to somebody who is struggling to understand, the simultaneity of speaking and listening is held in abeyance until the listener's delayed reaction catches up. Or when something feels funny when I hit my forehand in a tennis match, I know that within a couple of seconds, I will retroactively match up an internal injury simultaneous with the moment I planted my foot. Or to use Saussure's example, even if the 8:00 bus is delayed, it is still the 8:00 bus. Whenever a process is not at its stopping-point, when it has *not* happened at the same time it was supposed to, it is in its delay state, its secondary tempo. To say that something is delayed both refers back to its proper, virtual stopping-point and also describes its wanderings.

Delay occurs when a single event extends away from its starting point yet is also held back from departing. Delay prevents events from being instantly determinate, and only on that condition is temporal continuity possible. Delay breaks up the singularity of a simultaneous moment, but it keeps successive events tied to their common departure points. Delay is extended temporality held in place; it is divergent temporality's backstep into convergence.

(b) Delay in Husserl, Merleau-Ponty and Heidegger

Husserl

If simultaneity is a retroactive coordination of delayed reactions, how can memory preserve successive simultaneities?

If the mind somehow had a grid for containing memories, each experience could be indexed to a temporal counter. The mind's counter would start at birth to tick off dates, so that if something happened at the age of six and three quarters, it would be indexed to that number, and everything else that happened at the same time would be tagged by the same date. That would explain how simultaneities are constructed and recalled. But there is no phenomenological evidence for such a grid.

The succession of experiences, and hence the succession of simultaneous moments and the delay in their coordination, begins with a distinction between perceptions of the present and memories of the past. We might think of perceptions as clear, high-intensity images and memories as vague, low-intensity images. But Husserl points out that many memories are as intense as perceptions and that at any rate, intensity does not imply temporal location. The difference between memory and perception is not one of intensity but temporal intentionality.

What makes an experience count as a perception of the present is that I do not know how my expectations will turn out. What makes an experience count as a memory of the past is that I have already experienced the fulfilment of at least some of its expectations. Whether I got what I expected or not, some result already came about. The sequence of fulfilments of the memory ends in the present. A memory's futures are now contained in experience; a present perception's futures are not. The memory's future is also a memory. In short, the memory's future is an experience; the present's future is not an experience.

S. 24, 'Protentions in Recollections', begins like this: 'Every active memory contains intentions of expectation whose fulfillment leads to the present' (52).

The problem starts here. Since every experience as I have it contains expectations, the memory that retains that experience should retain its expectations. If I should seem to remember an experience with no expectations in it, I would not in fact be remembering the experience itself. If I seem to remember asking someone on a date, but my memory does not carry expectation with it, it would not be *that* experience that I was remembering.

Of course, some memories are of experiences whose results I still do not know. Nevertheless, all memory, even the most recent, has had some of its expectation stream carried out. Sometimes, hearing a melody through to the end changes one's memory of the individual notes at the beginning of the sequence. Indeed, in close-up, parts of the now lag behind other parts. In any field given now, I will already have noticed more of some elements than others. Strictly speaking, some of the now *has* already been fulfilled even now. Some of the now is earlier than other parts of the now. At this micro-level, some of a memory will always be unfulfilled, since some of a present memory of the past experience will not yet have arrived at the present as much as other parts of it. These issues concern the gradual emergence of an experience into the now, but they do not yet create a problem for whether memory calls up the past.

The next sentence in the text adds difficulty.

> Every primordially constitutive process is animated by protentions that constitute and intercept (*auffangen*) what is coming, as such, in order to bring it to fulfilment (76).

Each experience predicts (fortunately, not too accurately) what we will and will not notice, and what counts as its fulfilment. Protentions throw a net to catch something in the future. The present experience intercepts a future that has not happened yet. A memory also anticipates, or intercepts, its future, but the memory's future is already past. A particular memory picks out some of the past as its future, and excludes the rest of the past from counting as its continuation. Each memory experience thus prevents some of the past from passing into the present through it. To put it differently, each memory delays other past futures from arriving in the present. In other words, a memory carries into the present both what happened at a certain time in the past as well as the relevant continuations and fulfilments that extended the event from that past moment up until the present when it is remembered.

This raises the problem. If what gets carried into the present are the fulfilled elements in the original experience, then how is it possible for memory to reproduce the unfulfilled character of the original experience? In practice, we do often remember the original uncertainty in previous experiences. But we also diminish that original uncertainty by remembering the events in which it was resolved. To have an accurate memory that blocks the original experience's future, which by now we have experienced, from mis-colouring it, some memories have to be bracketed out.

Without this trick of blocking the fulfilment of expectations, to prevent the past's future from being experienced as already past, we would all become Kierkegaard's 'unhappiest man'. Kierkegaard says that it is a mark of unhappiness to lose what one has had in the past, and it is the mark of the greatest unhappiness to lose the past itself. The greatest disappointment arises 'not because the realization of one's hope is postponed, but because [the realization of hope] is already past and gone'. 'One remembers that for which he ought to hope; for the future he has already anticipated in thought, in thought already experienced it, and this experience he now remembers, instead of hoping for it'.[8]

Bracketing is necessary to delay the future that the past experience had. Yet, we have to know that the experience did already have a future, otherwise we would not mark the experience as a memory. To preserve a memory is precisely to delay, but not to cancel, the successors of the target experience in the past from having occurred in the present. To retrieve the past as past, the future of the memory has to get put off into the future of the present. A fulfilled protention must be treated as if it were not fulfilled. Yet as Husserl says, lines of continuation are continuous, so each of the subsequent retained moments *should* be included in each. Pragmatically, one might want to advise: When you remember an event, just don't focus on the consequences it had. But structurally, the consequences are part of the retention.

The future is in a sense closer to being present than the past is:

> In the intuitive presentation of a future event, I now have intuitively the productive 'image' of a process that runs off reproductively. Joined thereto are indeterminate intentions of the future and of the past, that is, intentions which from the beginning of the process affect the temporal surroundings which terminate in the living now. To that extent, expectational intuition is an inverted memory intuition, for the now-intentions do not go 'before' the process but follow after it (56).

In this first description, memory and expectation have the same structure, but reversed. I anticipate what I will find when I look deeper into the past, just as I anticipate the future.

> As empty environmental intentions, they lie 'in the opposite direction'. How do matters stand now with the mode of givenness of the process itself? Does it make an essential difference that in memory the content of the process is determinate? (56)

Does it make a difference that in memory we look for some specific experience, whereas futural expectations are more indeterminate? Husserl hesitates, since sometimes memory does not find anything determinate, only fuzziness. Conversely, sometimes when we look into the future, we are clear and certain. Husserl even says,

> In principle, a prophetic consciousness... is conceivable, one in which each character of the expectation, of the coming into being, stands before our eyes, as, for example, when we have a precisely determined plan and, intuitively imagining what is planned, accept it lock, stock, and barrel, so to speak, as future reality (56).

Just as memory can be as clear as perception, so expectation can have as precise a target as memory. Nevertheless:

> Expectation finds its fulfillment in a perception. It belongs to the essence of the expected that it is about-to-be-perceived. It is thereby evident that if what is expected makes its appearance, i.e., becomes something present, the expectational situation itself has gone by. If the future has become the present, then the present has changed into the relative past. (80)

The key to expectation is that it is about to be perceived, and when it is, the future has become the present. Since the past has already been fulfilled and cannot be fulfilled again, the fulfilment of a past event can only be another past; it cannot be the present because the past has already had its present. The reason why the attempt to remember the past *as* an expectation can never be accomplished is that it has already been fulfilled. The moment at which the past is fulfilled is, by the time the present rolls around, no longer in the future. We cannot get to the future of the past; we cannot wait out the past again. The attempt to regain it is infinitely delayed. Precisely because the past is given with certainty as fact, but cannot be relived, it is fact in the form of distance; it is 'a given', but not given, in experience.

Since the future is still about to be fulfilled in a present, the future is more present than the past. At first, the situation looks the other way around, since the past really happened. But the originary past is what is delayed infinitely, whereas the future can still be present; its delay can be resolved. Paradoxically, the past is more delayed than the future is.

Let us summarize the two problems of time in Husserl. The problem of simultaneity begins with the breakdown of a single Now into many object-directed time-lines; divergent stopping-points have to be synchronized; the solution requires a paradoxical staggered simultaneity and a backstep into convergence. The problem of the retention of anticipation is that the past's future has to be forgotten in order to remember the present's past; the solution requires a paradoxical delay of the already-given. That simultaneity should be distributed across lines of delay, and that remembered anticipations should carry with them a temporal line other than the line of presents, are not anomalous, but paradigmatic phenomena of temporal difference. In the context of Husserl's phenomenology, these patterns come up as problems for a theory of temporal experience. But articulating the problems in this way leaves an impression of Derridian deconstruction. In fact, in Chapter 7 what I will call Derrida's solutions is close to what Husserl anticipated in creating these problems: the simultaneity of the non-simultaneous; the now's return from the quasi-temporal future; the non-occurrence that drives the time-flow; delayed protention, and indeed, delayed retention; and the economy of time-squandering. These are not so much problems for the philosophy of time as descriptions of the structure of time.

The simultaneity of an experience with its anticipations can be captured in memory (as Husserl puts it: simultaneous experiences in the present pass into memory at the same rate) only if one of the simultaneous elements is in delay. It is therefore in the nature of events, as intentional objects, to pass into the future in staggered deferrals. Exactly how deep delay goes becomes a central question for phenomenology. Husserl never quite drew the full conclusion; Merleau-Ponty draws it fully at times, only to draw back; Heidegger entertains so much delay that time slides out of succession almost entirely. The complete phenomenological implications await philosophers on the outer edge of phenomenology, like Derrida and Deleuze.

Merleau-Ponty and Heidegger on delay

Merleau-Ponty analyzes cases where the multiplicity of anticipations is stretched so far that a single ego-subject cannot synthesize all the expectations that an interpretation of her world requires. Heidegger analyzes cases where it becomes impossible to fulfil expectations, and the subject can only hit the pause button on experience and wallow in delay. The danger with Heidegger is that the overstretched time-line gets taken out of time; the opportunity is that delay might reveal a new field of experience with a new temporal structure.

Merleau-Ponty on two schemata of time

Prior to the 'Temporality' chapter of the *Phenomenology of Perception*[9], Merleau-Ponty asks how sense is bestowed on an event from one's past. Two accounts are considered. The first thinks of the past as 'prolonged infancy' (*enfance prolongée*, 398), which, in due time, has effects that throw light on the past's true nature. Merleau-Ponty calls this 'artificial' (*factice*); it undervalues the intensity of the earlier event as it was experienced at the time and overvalues retrospection. The second alternative says that the past remains available in its own terms for future sense-giving, and hence that full retrospection is always 'deferred' (*differée*). In this preferred alternative, the sense of events is deferred through succession (rather than prolonged through progression). Time is deferred sense.[10]

The 'Temporality' chapter offers a third, still more preferable, account: 'It is in one single movement that, from one end to the other, time sets itself in motion. The "instants" A, B, C, do not *exist* successively; [together] they *differentiate themselves* (*se différencient*) from one another' (479). Time surges up all at once (*toute à la fois, toute entière*), once and for all (*une fois pour toutes*), at one stroke (*à un seul coup*).'

How do we reconcile the second and third accounts? The second says that an event enters history by deferral; the third says that it enters history by simultaneous differentiation. How do we reconcile *différer* and *se différencier*?

Merleau-Ponty's 'Temporality' chapter deals with many questions and is difficult to map. To order my exposition, I resort to a Kantian rule. When in doubt, analyze temporal schemata under the categories of quantity, quality, relation, and modality as time-series, time-fullness, time-order, and time-scope, respectively.[11]

In general, temporal schematism presupposes a subject. Time is like a river in that it is a single 'permanent' series (482); we pre-reflectively feel its unity without having to attend to every single wave or moment. But time is not inanimate. Without a subject who compares events, there is no past, only present facticity. Even if there could be a past without a subject, it would contain no direction, only coexistent 'eternity' (470–1). In short, time without subject would have no now, no past, and no order, only flat eternity. However, the type of subjectivity necessary for, and permitted by, time will prove problematic.

Time-series: Kant says that to schematize a time-series, a subject attends to a point, imagines another point next to it, then cognitively runs through a plurality and holds them together as a totality. Merleau-Ponty's spin on this is that in order to produce the second point, one must already 'feel the pressure' of the future in the first point. For a present to have a successor, it must already feel like the past (473). Past events feel past because they feel like they led up to something. Merleau-Ponty draws examples from his own experience: his memory knows where to date the visit he once made to Corsica, because that visit included the foreboding of the war to come. The futural pressure in the event locates it in a time-series. This is the interplay of deferral and simultaneity. An event is present relative to the delay of some of its parts. By holding those part-events 'at a distance', we survey the time-series. Events at the point of presence are deferred to the next point, and because of this, the total series can appear simultaneously.

Time-fullness: For Kant, the quality of one and the same experience can increase or decrease in intensity over time; this measures the fullness of time by degree and is essential for temporal continuity. Merleau-Ponty uses this to explain how a present event retains the same content once it is past. Often, he says, we think that when a present experience A is replaced by B, A is turned into a new experience: A^1. And we think that once B is replaced by C, A^1 is replaced again by A^2, and so on (477–80). In other words, we often think that when an experience becomes a memory, its content changes. But then the A's would not be identified. Memory would seem to jump across unexplained gaps, and events would have little to do with each other. It would feel like our entire set of memories were replaced every few seconds. It must be wrong to distinguish A-present and A-past by content. Better to think of an event as a pebble in the riverbed that we can still see, even though many waves wash over it. Attending to A-past means precisely to attend to the fullness of A, to recapture the hopes and fears of the A-event. By this means, memory 'gradually links us to our whole actual past'. The significance of an event, like the meaning of a sentence, survives all sorts of changing experiences. Of course, we can forget things by degrees down to zero. But just as the time-*series* posits the whole of time by deferring parts of it, so time-*fullness* retains the whole of time to the extent that it delays the decay in intensity of its parts. Time-series delays successors, time-fullness delays decay in predecessors, and together they maintain the continuity of time.

Time-order: Kant's example of ordered succession is a boat moving irreversibly down a river. To experience succession is to understand why the order of our perceptions makes causal sense. If the boat appeared at random places in successive moments, we would not experience time-order, and our understanding would collapse. Of course, some things can be experienced in any order whatever, like the rooms in a house, but those things do not exist in succession. They are all there at once, simultaneously. Merleau-Ponty's paradigm of time-order is similar, but his example is the narrative of one's own actions. Action makes sense only if it draws possibilities out of a personal history, actualizing one step of a project at a time. The order of not-yet, now, already, and still to come, is irreversible. The present is a *starting*-point or 'centre' (489) and the future is the vector that it gets under way. This is why the order 'maturing' out of a single moment, constitutes a 'whole destiny', 'at one stroke' (486). For Merleau-Ponty, time-order is not just serial moments in a passing present, but a totality running from the beginning to the end of a meaningful process. Kant had described a time-series by the way it 'runs through and holds together' a set of points, but for Merleau-Ponty, the beginning and end of a narrative determines the order of the intermediate steps, and makes a totality or destiny out of the process. What is controversial about this, is not so much Merleau-Ponty's hyperbolic statement that 'we are the upsurge of time' (489), or that when we put our meanings in order, 'time knows itself' (487). What is controversial is the implication of 'destiny'. Merleau-Ponty cites Swann's love for Odette, which is inseparable from his subsequent jealousy; it is because the former 'provokes' the latter that we say it happened first. Swann's love 'already' had a jealousy-'manner'. No matter how opaque that was to Swann himself, his whole destiny was ordered into the starting-point (486). Now, in the case of the fictional character Swann, this seems true. But how much of me must there be in a moment of time? How little freedom do I have once the upsurge of time is me? The topic of freedom leads naturally to modality.

Time-scope: For Kant, a possible event can happen at various times; an actual event happens at a particular time; a necessary event happens at all times. Together, these modalities ensure that there is a single time-line, though it contains many actual and possible event-sequences. The singularity of the time-line is based on the necessity of the potentiality of actuality. Whereas time-*order* depends on our narrative centres; time-*scope*, which covers history before we were born and after we die, must not 'start from a central I' (476). My experience of it departs from my own 'perceptual field', but it travels along 'horizons' and 'backgrounds', picks up intensities of 'weight' and 'style', 'traversing' all events no matter what the order of their dates, making each event 'revert' to a single field of signification. As in Kant, time-scope synthesizes all the other schemata, making good on their deferrals. 'With every moment that arrives, the preceding moment undergoes a modification' (though not a change in content) (476). The present does not just push its predecessor into the past; it puts it on the map. This is the thesis of simultaneity that we started with, which is in the end compatible with deferral, namely that, 'It is in one single movement that, from one end to the other, time sets itself in motion'. The schema of time-scope brings even the most distant possibilities back into my time-line.

This existentialist spin on transcendental schematism makes an interesting theory of time and offers good material for simultaneity and delay, in spite of raising flags about destiny and me. But it is only half of Merleau-Ponty's story, because so far, we have considered schematism under what Kant calls transcendental analytic, not yet under transcendental dialectic.

For Kant, 'reason' is 'dialectical' in that it illegitimately conflates concepts and intuitions, as if intellect alone could tell us what is in the empirical world, or as if imagination could create a world beyond empirical phenomena. Now, Merleau-Ponty rejects Kant's strong separation between universal categories and particular intuitions. So whereas Kant warns us not to accept dialectical versions of the schematisms of time, we might expect Merleau-Ponty to extend schematic time into dialectical forms. And in fact, for each phenomenologically common sense analytical thesis that we have seen so far, Merleau-Ponty articulates an exorbitant, dialectical, proto-Deleuzian, variation. That he returns to normal on each occasion does not necessarily mean that he is living in the past. The tension between normality and excess may be symptomatic of the relation between simultaneity and delay.

Dialectical time-series—the mobile milieu: The time-series moves throughout its whole length as soon as any moment arrives: it is a 'mobile milieu' (*milieu mouvant*, 480). It is one thing to say that a given moment of an event makes sense only in relation to the whole time span of the event. But it is worth pausing at the strangeness of this (Bergsonian) picture. When one event occurs in the present, does every event in its history start happening again? Is the whole time-series just one event that moves by waves into past occurrences and future possibilities, extending the present in all directions at once? Merleau-Ponty wants moments in time to remain distinct. Yet he says they 'run into each other' (481), and each is 'just one aspect of the total bursting forth or dehiscence' (480). This is the dialectical extension of the time-series schema. Kant's analytic distinguishes intuition from concepts, unit from unity; Merleau-Ponty's dialectic asserts that the unity of time is intuited as a unit of meaning.

Dialectical full intensity—free travel: Merleau-Ponty is concerned that if all time is immanent, and past and future are as near to us as the present, then we would be able to 'travel freely' (*cheminer librement*, 474) through time as we do through space. But then the transience of time would be lost. 'There is only time if it is not completely deployed, if past, present, and future *are* not in the same sense' (474). Merleau-Ponty does affirm 'a sort of eternity' (475), so he only half resists time travel. But if the synthesis of travel and transience depends on the incomplete fullness of events at a given time, one wonders whether Merleau-Ponty's examples are not too full of themselves. War looming in Corsica, Swann's jealousy, a river. Large desire, large geography, large politics. Why must time be like a river, and not like sitting on the dock of the bay wasting time? Why appeal to one large direction of flow, rather than to a vague drift, or to an artificially structured outline? If we break up the large examples, could we accept the dialectic of consciousness travelling freely through its history and resist the safety of each time in its place? Objects that are more assemblage-like might have complexities of rhythm and feedback loops whose investigation looks more like time travel.

Dialectical time-order—someone mythical: Analytic time-order for Merleau-Ponty is a person sliding into jealousy. Merleau-Ponty emphasizes the 'cohesion of a life', equally present at this instant and during 'my whole life' (481). In '*my* life', he says, 'I am myself time' (*je suis moi-meme le temps*, 481)—it is not just that time exists *for* me, but that time *is* the ordered series of my life, the 'interior communication' of my experiences through time. But in the same passage, Merleau-Ponty appeals dialectically to 'mythical personifications' of Time, the feeling that 'time is someone' (*le temps est quelqu'un*, 482). While the latter seems merely to generalize the motto that 'I am myself time', it is quite different, replacing the possessive pronoun with an impersonal, fictional subject. Grounding time-order in 'me' and my destiny is different from grounding it in a depersonalized, decentralized field of significance. Appealing dialectically to a mythological person named Time does not in fact identify time and subjectivity as closely as appealing to 'me myself' does. 'My time is me' limits time-order to one man's life, generalized under one character's narrative. 'This Time is someone', in contrast, incorporates into time-order many unsynchronized destinies, so that 'my' memory includes memories, anticipations, decisions, and destinies of some anybody, that is, of others. Again, Merleau-Ponty's text is undecided: the dialectic of anyone-ordered-time and the flux-subject coexists in his descriptions alongside the analytic of me-ordered-time and the destiny-subject.

Dialectical time-scope—tempos of quasi-eternity: Time-scope covers all possible times, but is 'actualized only bit by bit' (*ne s'effectuer que peu à peu*, 483). 'Bit by bit' does not just mean one instant at a time; it means that each event and each person has its own manner of setting the totality of time in motion. Each present has its proper 'tempo' (*son tempo propre*, 483). Each event delays its successor for the amount of time appropriate. Each has its own way of speeding up or delaying its resolution and its own way of reopening its predecessors. Since nothing is lost in the whole, this 'grounds quasi-eternity in the event' (*fonder cette quasi-éternité sur l'évènement*, 484). Each event's tempo determines when other events around it can start sounding, and those events in turn have their own tempos, which determine events around them, and make the first resonate around it again. Founding the total time-scope on actualized events yields a kind of eternity, self-differentiating rather than naturally coexisting. But it is still only a 'quasi-eternity', since dialectically, singular events become their own universals. The analytic schema describes events that 'reopen' into temporally located pasts and futures; the dialectical schema describes quasi-eternal dislocated events in which 'time re-starts itself' (*le temps se recommence*, 484). As always, when Merleau-Ponty departs from critical philosophy in favour of dialectic—in favour of the mobile milieu, free travel, the mythical someone, and the quasi-eternal tempo—he hesitates. He worries that 'cyclical rhythm' (484) might create the 'illusion' that eternity could be grasped all at once. Just because time is not like a river, it should not mean that it is like a house. This worry makes him vacillate back from rhythm to destiny.

But once multiple tempos are introduced into temporality, the consequences are unstoppable. Once each event restarts others according to its own tempo, then each event is moving at different rates of speed at the same time. To say it radically, once fine-grained polyrhythm is introduced, 'temporality' might turn out not to be a very good concept. It might be that 'tempo' is not a sub-topic but an alternative to the concept of time. At the very least, there is no one river of time but coexisting micro-streams, corresponding not to large objects like jealousy and war but to micro-desires and micro-politics. Indeed, Merleau-Ponty's penultimate sentence in the chapter hints at 'collective history' (495) rather than 'my life'.

In sum, Merleau-Ponty schematizes both analytic phenomenology, where deferrals of intentionality are resolved in the time-scope of my life, and dialectical phenomenology, where self-differentiating deferrals add up to a vague someone belonging to quasi-eternal poly-tempos. There is often only a fine line between these two strains in Merleau-Ponty, but they lead in different directions. Perhaps one of them is the true Merleau-Ponty. But perhaps the two temporal schemata truly coexist alongside one another, and there are two coexisting types of temporality.

If we posit two kinds of time, though, we risk a cosmological split between transitory and eternal time. The thesis of this book is that delays and coexistence, transience and simultaneity, are inextricable. Nevertheless, delay has a life of its own that is not always brought back to the simultaneity of over-viewing destiny. However, if delay is so pervasive that there is no point hoping to complete projects, as Heidegger sometimes envisages, then getting into the existential mood of delay will function much like a simultaneous overview of time, as if to take an eternal point of view. But unlike eternity, universal delay will feel like transcendental waiting in limbo.

Heidegger on waiting[12]

Bernard Waldenfels surveys time lag effects in phenomenology beyond the topic of perception.[13] For example, the material that is in other people's consciousness is defined by what I do not yet have in my consciousness—the other's is my delayed content.[14] Part of being in time is the capacity for indifference towards the particular moment. This explains why time is experienced as a threat, even while it expands horizons for engagement.

Engagement introduces new issues of simultaneity and delay. In *Being and Time*[15], delay is longer limited to the partiality of epistemic fulfilment. For Heidegger, I experience the world in a short allotment of time I cannot control. Dasein is thrown into history and space, but uncertain what that means. Dasein is not here, but out there, standing-outside present selves, in an 'ek-statical unity of time' (350), not in the now, but already in the future, in the 'raptures of the future, past, and present' (350). We are never *just* in the present, or just in the past, or just in the future. The whole notion of expectation is that we are thrown forward, outside of time just when we are in it. When Heidegger posits the future as the first moment in time (376), the role for delay in past, present, and future alike is intensified.

If action were by and large successful, the future would be predictable; but insofar as action is uncertain, the future is awaited. The sections of *Being and Time* on waiting concern time and care. To care about something is not just to know what it is and plan for the future; it is to take oneself as being involved in the life of something. Heidegger is a transcendental communitarian environmentalist: one cannot fail to care about the future of others. Caring is involvement ahead-of-oneself.

Heidegger describes three kinds of waiting. The first concerns things in their normal course to completion. The second concerns things breaking down and held up. The third is when the breakdown of expectation makes us sit down and think. Heidegger tries to show, against Husserl, that cognition arises *after* normal anticipation-fulfilment chains break down.[16] Delay is good; when concrete involvement breaks down, we can speculate.

In the first kind of waiting, a person wakes up in the morning, and whacks the ready-to-hand alarm clock without delay. But bodily immediacy does not happen in isolation. Practical experience is not a series of handles and buttons; it is a network. Being involved *in* something implies getting *to* something else (353): preparatory moves, estimated times of arrival, follow-ups and bigger projects. Involvement entails delay. No one can be aware of all the relevant anticipations in their own activities. There are relevant anticipations about one's own future that one cannot have. In spatial terms, the world is spreading out. In temporal terms, the now already includes things that happen in the not now. Part of the now is opaque. It is not just that at each now, the future is delayed. Because the future is, in extra-conscious anticipation, already inside the now, the delay is *in* the now, not just the delay *of* the next but also delay within the now.

As soon as there are fuzzy expectations about when exactly something is going to happen, and the world unfolds to partial events and multiple tempos, delay is part of the structure of time. When we expect the plane to arrive at a certain time, in one sense we expect just that; but when we take into account the time when we will receive our bags, or the time when that awful person, who makes us wish the delay were even longer, will come to meet us at the gate, and the times when all the other sub-events might occur, then some aspect of every event is going to be delayed. The normal anticipation scheme's complexity creates delay.

Waiting aims at the future, since everything in time takes time; but it aims from the past, since a person has always already carried out preliminaries. In being involved, one 'comes back to oneself' (353). Waiting is virtually the definition of the lived present. (Heidegger puns on *gewärtig*, 'awaited', and *gegenwärtig*, 'present'.) Waiting means letting beings take their time to be.

A second kind of waiting occurs when the predictability of after-effects breaks down. In this form, 'waiting is neither concerned with the goal or the expectation of the impending finish…, letting something be involved is constituted rather in the unity of a retention which awaits' (353). Here, since waiting is openness to the present, we have to be willing to deactivate or 'forget' our retentions (405), and therefore to forget ourselves.

This connection of 'retention' with 'forgetting' rather than memory (354), which would be an oxymoron to Husserl, consists of expecting the past to continue but without thinking we know what we have been looking at. Heidegger's examples involve skills that are neither conscious nor unconscious. On a wilderness hike, a beginner like me who does not 'know my way around' cannot rely on anticipation. I have to wait for the next stone to show my foot where to go. This kind of waiting allows something other than memory to control the retention of a goal.

If we thus detach waiting from anticipating, yet retain the principle of involvement, we come to the tool whose effect is unpredicted. The hammer wobbles and I have ruined my house. When things do not work, the future is 'held up'. A 'diversion' or 'disturbance' occurs, and attempts to eliminate the disturbance do not succeed. Normal waiting, as when we glue something and wait for it to dry, is a determinate waiting for things to 'temporally run their course' (355), whether on schedule or a bit late. But when the glue is not sticking, we have to wait either extra-determinately, staring at the obstinate materials, or under-determinately, off into space.

When involvement fails, we have to admit that we cannot do this work today. Tomorrow we will hire an expert. Who knows how long that will take?[17] Maybe it is just not going to work. The project comes to a halt. We become 'absorbed' *in* the waiting, as if the future we are waiting for is suddenly not ours. When our activities are 'held up' (*aufhalten*), our hold (*halten*) on things is suddenly not *our* hold. There was going to be a time lag, but now we do not even get to experience the delay as part of our living present. What has been delayed is the delay itself. When you spend a year in France, you need an official doctor's appointment to confirm your residency status. The *préfecture de police* tells you that 'your doctor's appointment is at 3:00'. But you know the bureaucracy, and so you ask, 'will I actually see the doctor at 3:00?' and they say 'no, there is a one hour delay'. Your 3:00 appointment is for the time that you are scheduled to start waiting. You might think that it would not matter if you come later, at 3:30, since the doctor will not see you for another half hour. But 'no, you didn't come in time to start waiting'. And if you come at 2:30, 'no, come back at 3:00, you still have some waiting to do before you can start waiting'.

When waiting is in limbo like this, as in boredom or insomnia[18], we are placed in a 'deficient present' (355). Sometimes it is still worse, and the 'resistance, hindrance, and disturbance' are insurmountable. All we can do is hold 'aloof' (356): There is nothing we can do but wait until things get better and then make a new plan:

> The temporal structure of resigning oneself to something lies in a *non-retaining* that awaitingly makes present... (407).

We obviously cannot cling to a plan that is not working. Giving up becomes a new kind of memory and a new kind of waiting. In contrast to retention without memory, which I characterized as following through without assuming we know what has been happening, memory without retention means evaluating what has been happening without hoping to follow through. What is the status of delay when we give up?

When a radical delay breaks down a predictable delay, one response is to try to repair the time-line. But in another response, the third kind of waiting, we ask ourselves what made us think we could master time in the first place. At the moment where the hammer does not work, we might ask, 'why not move to an apartment?' or 'what does it matter, we're all going to die', or 'why did Western civilization block itself in with houses?' Any small breakdown of expectation is an opportunity to stand back from the temporal advance and 'survey' (359) once again, as if from scratch, in crisis or aporia, the situation as a whole. When your house or your marriage is breaking down, hit the pause button, go somewhere else, think again. In the second kind of waiting, even if you have no idea what comes next, you are still involved in forward-looking anxiety. By contrast, hitting the pause button brackets out the temporality of the world. This does not cancel out interest in the event, but it lets you deal with the material element of the event, or the conceptual element, or the intersubjective element, without the temporal element. Of course, we do not literally pause time; this third kind of delay has its own quasi-temporal structure in the form, 'just-looking-around' (357). It involves memory (survey of the past) and retention (following up on the past) but not lived time. Frozen simultaneity is inserted into the temporal flow, and delay becomes its experiential field.

Heidegger describes several stages in this transition from practical to speculative waiting (357–9). In the middle of a job, we can 'tarry' while we 'consider' our next move (358), a contemplative waiting within practice. When one module of a job is done, and things are at a 'standstill', there is time for 'inspection' and 'checking up' (358). When 'surveying' an open possibility, there is time for 'deliberation', for 'envisaging' (359), even for existential analysis, as long as theory delays practice, and seeing delays touching.

Delay is a necessary and not just a contingent category for a question-asking being. Without the final, existential upshot of delay, a person might experience one part of an event taking longer to complete than another, but would never feel the differential as a lag. They could say, 'I can't wait for lunch', but lunch would be on schedule. They could experience subjective delays, desiring things before getting them, and they might experience delay as a severe ontic-objective problem, but they would not experience delay as an ontological rupture, and would never need to hit the pause button and think again. For Dasein like us, the feeling of delay is not just a gap between two normal, actual events, but a gap between the virtual, non-normal event and the actual, normal event. The full sense of delay occurs with the gap-conferring relation between the practical functioning of the world and the existential anxiety of the surveying question.

In sum, things can be continually interactive with predicable delays, discontinuously interactive with delayed delays, or non-interactive, where quasi-atemporal surveillance makes an experience out of delay itself. Delays fall under schemata of anticipation, disturbance, and pause; the delay we know (frustration), the delay we know we do not know (alarm), and the delay we do not know we do not know (anxiety); delay in time, delay in being and time, and quasi-non-temporal delay.

It is hard to be sure whether to use the word 'delay' for all of these cases, or even if delay is a unified phenomenon. If it normally takes three strikes of a hammer to drive a nail in, is the paradigm delay found in the time after the second strike while waiting for the third, so that delay is the time-lag built into a schedule? Or is there only delay in the rare case where the third strike does not finish the job? And is there really delay if there is no specific expectation? After all, delay seems to require that the anticipated experience be prevented from arriving, not that some other experience be substituted for it. If you are supposed to finish your MA thesis in five semesters, and are now in your seventh and getting nowhere, and you take a leave of absence to rethink your options, are you delaying your graduation or not? The university counts the semesters you are officially registered in, that is, paying for. As far as they are concerned, a leave of absence is a pause but not a delay. By definition, if you do not pay, you cannot be procrastinating. Perhaps the bureaucracy is correct. For something to be delayed, it has to be *that* thing that is being strung along. A delayed event must not only be distant from the now, it must be tied to the now elastically, emitted from the present yet spread out into the future. Delay concerns the same event continued into the future (otherwise there is a change of project and not a delay in the original project) and yet strung along beyond the future (otherwise the project is completed on schedule). Delay is an event on a string, already emitted from the present and retained from the past, yet whose past and present are interrupted by already having been spread into the future. As the action continues, it pulls the event along; or to say it differently, the event is attracted backwards. Yet since some elements of the event are not able to occur now, its fulfilment is repelled by the present.

The question is whether Heideggerian existential delays surpass, or remain tied to, Husserlian delays of everyday experience. It seems too abstract to say that delay calls an indefinite halt to the struggle for concrete temporal prolongation, that is, that existential delay delays normal delays indefinitely. I think, with Husserl, that anticipation schemes are inherent in delay, and conversely. For Husserl, we resolve the little crises of beings within experience, rather than by Heidegger's lateral move to Being. For Husserl, the issues of delay are: what is coming next, how do we calculate, what are the rhythms? For Heidegger, the commandment of delay is: hit pause; bypass the ontic time of clocks. In short, Heidegger runs the risk of shifting the temporal problem of delay into atemporal ontology. Deciding between Husserl and Heidegger means deciding whether the founding delay is determinate or aporetic; that is, whether delay is engaged or reticent; that is, whether delay synchronizes and de-synchronizes to and from simultaneity, or becomes its own region of *totum simul* ontology. In the chapters that follow, Husserl more than Heidegger will lead us.

Still, Heidegger leaves us with a good question: can a total break with anticipation create a new kind of temporal assemblage and a new field of delay-temporality to analyze. The delay experience might go beyond both Husserl and Heidegger, both beyond the next event not happening when it should, and also beyond post-temporal questioning, and be a way (as in Deleuze) of experiencing the world through the very elasticity of time. Can Heidegger's breakdown of anticipation, with its overlay of normal and disturbed intentionalities, generate new experience *in* the delay, without bypassing ontic concern? Can radical delay generate concrete new events? Can delaying an event make it simultaneous with another event at a distant time, both striating and smoothing out the time-line, radicalizing events without pausing the time-line indefinitely?

Of course, theological perspectives likewise hold that delay pulls us beyond normal anticipation towards eternal realities. According to theology many people think they can be fulfilled in time, but only because what they think of as their goals, fulfilled or deferred, do not pertain to their spiritual lives. People's true ends are delayed even when they experience no temporal lack.

Before investigating what kind of temporal field delay initiates, we will look further into ideal simultaneity. This takes us to the relation between time and eternity in Ancient and Medieval philosophy.

Chapter 3

Simultaneity and Delay in the Era of Eternity: Ancient and Medieval Issues

Theme of the chapter: Simultaneity synchronizes the temporal with the eternal. The resulting teleology is a state of delayed gratification.

(a) Simultaneity as Synchronization: Plato and Plotinus

For Plato and Plotinus, eternity (*aion*) is perfect being outside of time (*chronos*), all events together simultaneously (Greek *hama*, Latin *simul*). Eternity can play the role of total simultaneity, paradigm simultaneity or the only genuine case of simultaneity, the cause of simultaneity or the life of simultaneity. On the surface, simultaneity and eternity seem opposed: simultaneity suggests co-presence for an instant, whereas eternity suggests constancy over time. But eternity can mean something that exists all the way through time, coexisting with every transient instant. This might best be called 'all time', *hapantachronos*, sempiternity, or perpetuity. *Aevum*, the time of the angels, is a middle ground, eternity manifested through time.

In Plato's *Timaeus*[1], a stranger tells that the eternal forms are put into the reality of changing things by the demiurge. He mediates, by force or by persuasion, the pure world of being with the world of becoming.

Early in creation, the demiurge creates time: 'Time is the moving image of eternity'. 'Being in the image of a form' could mean that everything is somewhat like the eternal truths. Things that change would have something like eternity, to the extent that they endure at all. Alternatively, 'being in the image of' might mean that only the form is eternal and everything else falls into corrupt non-eternity. In the first, positive sense, time reveals what eternity is. In the second, negative sense, eternity alone is real and time is its untruthful shadow.

Plato ties the thesis that 'time is the moving image of eternity' to the movements of the planets.

> He planned to make a movable image of Eternity and, as He set in order the Heaven, of that Eternity which abides in unity He made an eternal image moving according to number, even that which we have named Time. For simultaneously (*hama*) with construction of the Heaven, He contrived the production of days and nights and months and years which existed not before the Heaven came into being (37d).

Days and nights came into existence only with the creation of the planets. Time did not exist before the concrete world.

> And these are all portions of time: the 'was', the 'shall be', and so on, it is wrong to apply them... to eternal being.

When we talk about eternity we should, with Parmenides, just say '*is*'. 'Was' and 'will be' are applicable only to becoming, in time.

To be in time is to be in motion, and in particular, to move according to number. When the demiurge made the bodies of time,

> He placed them in the orbits along which the revolution of the Other was moving, seven orbits for the seven bodies (38c).

The moon moves in a circle around the earth; the sun is the second above the earth. Plato gives a long story about the different planets and stars and where they are moving. In order for there to be time, there has to be movement or change in general, and in specific, regularly moving bodies. This is the difficulty. There are two levels of time. On the one hand, time is the concrete, physical movement of the planets. Time could not exist before the planets moved. Time is the movement of things, not just any old things but planets and stars, whose circular movement is perfect. Their circuits constitute pure 'numbers of time' (38c). There is no measure in other kinds of alteration. When things get hot and cold, for example, there is no guarantee that they do so regularly, at the same pace. In contrast, things that move in a circle are regulated, always the same. We can time them, and time other things by them. We cannot count on other kinds of change; they are uncountable. They do not have the same number of time; their numbers are bewildering, whereas the heavenly bodies are like pure living numbers, divine animals.

But the other level in this theory is that it is numbers that make time. The second level of time is not concrete movement, no matter how perfect, but abstract measure.

These two levels make it difficult to interpret the passage:

> Simultaneously with the construction of the heavens, He contrived the production of days and nights and months and years… these are portions of Time… (37d).

The two structures of time, the concrete and the abstract, the bodies and the measure, exist at the same time (37e). The measure of time, which is the same always, and the movements of time, which always change, exist at the same time. Movement and the motionless occur at the same time. This is odd, since the various segments or 'parts' of time, which are not simultaneous with each other, all have to be simultaneous with time in general (37e). But there is no other way to put it. The flow of time and the measure of time are both what time is, so they have to coexist in the same time. At any time that the planets are in movement, there has to be a way of measuring time by number. And any time there is a way of measuring time by number, something is in movement.

But it is not enough to say there are two aspects of time without saying which side is prior in which way. Plato says the sun and moon came into existence to determine and preserve time. There is an important sense in which the movement of the planets *is* the numbering scheme for the clock of time. Their movement is the measure of time in other things. So in one sense, the abstract measure of time is prior, tied to eternity beyond time. In another sense, the concrete is prior, regular motion that synchronizes clocks.

It seems like a vicious circle. We know the planets are perfect because the clocks show they are regular, and we know our clocks are right because they are synchronized with the planets. The planets could speed up and we would never know it, since they are what we measure our clocks by. This seems like a problem, but how else could it be? The abstract and the concrete must be synchronized, and there is no third measure by which to synchronize them.

Once Plato gets to the planets and their movements (79f), the narrative gets complicated. There are movements of the same and movements of the different. The universe as a whole goes in one direction: the direction of the same. Within this moving space, the planets move in the opposite direction: the direction of the other or the different. Each planet and star moves at a different rate of speed in the direction of the different. The planets are less regular than the sun and moon, so Plato calls them errant wanderers. Not as easily seen, there is something tricky about their movements. All of the planets make circuits, but it is as though they are in a race. Because of their different orbits, which cross from time to time, it looks as though at times some planets catch up to others and overtake them, as though some go fast and then slow down and actually reverse course for a bit before they take off again. Calculating by the heavenly bodies is therefore complicated, because it is as though there are seven different clocks, or one clock with several rates of ticking. To make matters worse, we have to measure our own motion against these differences. Plato calls this the bewildering quality of time. Luckily, he says, every once in a while, in spite of all these strange patterns, things come back to a pattern we saw once before. These are the 'great' cycles, or 'complete years' (39d). During the course of a year, confusing things happen, but when the year starts again, everything is back in its starting position. It might not exactly be a year, and it might not be that every condition repeats at the same time, but there are cycles where we can say we are back at the beginning. Even if there is no strict beginning, every once in a while a motion occurs that has happened before. Many philosophers, as we will see, say that in order to stabilize time and know when we are, we have to return to something.

I draw three consequences. First, in Plato's story, time requires that there be not one but multiple speeds, directions, and counter movements to explain and triangulate given times. The way we measure two events happening at the same time is that one thing is moving quickly, and another thing, at another speed, slices into its trajectory. We cannot know what time something happened by looking at *one* moving thing. But if we can say, 'this event happened on earth at the time when Jupiter was aligned with Mars', there is a crossover point between two measures, so we can know that something happened here when *that* crossover happened. If all the planets were moving in tandem, their movement would be continuous and unmarked by events. Even simultaneous motion is therefore time on the order of the different. Discontinuous movement occurs in special moments when one body overtakes another, so that we can measure two speeds in relation to each other according to a third. Measurement requires a stopwatch to observe and control the stopping of time at the moment of overtaking. Of course, a stopwatch does not stop time, but Plato's word for 'overtaking' (*katalambano*) suggests stopping, control, holding steady, or arrest: controlling stopping-points and holding multiple speeds steady in thought.

The second consequence concerns the wandering of planets. Since time is measured differently by different objects, the motion of some things must not be regular but bewildering. It is not that time cannot be measured with precision. Each clock or each combination of clocks gives a precise measure. But there will always be different ways of measuring the same event, and there is no certainty over whether a certain triangulation of clocks is the right one to use. Indeed, this is a general problem for Platonism: there are many Forms by which to understand the world.

The third consequence is that the multiple measures of time must come back to shared eternity. Plato's story cannot work without a complete year to come back to, the moment at which succession is generated, which starts things off again. This must be able to happen in the actual world of planets, but it also appeals to the theory that time is the moving image of eternity: to the point of time when abstract time measure and concrete time motion line up. Separating time from diverging motions makes it easy to imagine universal simultaneity, but it makes it harder to see how to re-synchronize those motions once they get out of line.

But do abstract and concrete time line up according to abstract time or to concrete time. Which side explains the other? What is the common ground between eternity and time? Plotinus offers a synthesis: the eternal expressed in concrete time is life, or world soul.

Plotinus' treatise 'On Eternity and Time'[2] falls into three sections. The first proposes that if time is an image of eternity, we should be able to look at concrete changes and see properties analogous to eternity. The second section, from s. 7, asks whether we can explain time by starting with images. In the third, from s. 11, he reverses the question and asks whether we can explain time by starting with the eternal.

The first section dismisses several options. If time is the moving image of eternity, is eternity therefore in motion, that is, in time? No, by definition. Could eternity be a combination of motion and rest? No, because motion and rest are differences, and eternity is unity. Is eternity the unity of changing things, in the way that life is a unity of organisms (s. 3)? This is promising. A life is not an aggregate of things a person does; it is the person persisting in movement, a drive that expresses itself outwards, that produces and unifies events. Perhaps eternity is to time as a person's life is to their or her actions (s. 5).

The model of unified life allows Plotinus to maintain that eternity is a constant, not undermined by changes in what was and what will be. In a way, eternity is all that ever happens, but it is neither a series of discrete events in succession nor one event happening in perpetuity. Eternity is the always shining forth of what happens. Like a single point preparing to flow into a line (s. 3), it is the stored up energy out of which changes arise. That moment is timeless, but it is the moment *at which* time, 'without delay' (*oude mellei*, s. 4), pours forth into change. In Husserl's terms, it is the pre-simultaneity of succession: the capacity of a single point to represent its extensions in all directions.

In general, eternal Forms shine forth in individuals (s. 5). In particular, every object has a stable nature ready to flow forth into a series. A living thing is all there with itself, a local eternity, 'already' what it is (s. 5), simultaneous with itself in advance. Eternity means simultaneity with itself, but simultaneity is not static. It continues to shine forth from within, for the time it takes. Every reality is in its own way eternal.

In s. 7, Plotinus says we need to understand how actual temporal changes work. We need to move down to time, to see it from below. He returns to Plato's question concerning what sort of clock measures motion. Is time the same as motion? No, motion is *in* time. Is time measured by the distance of motion? No, different objects move different distances but all do so in time. If we measured time by distance moved, there would be 'many times at *the same time*' (*polloi chronoi hama,* s. 8). From Aristotle to Bergson, philosophers tend to reject the thesis of many times (perhaps rashly).

Finally, is time the numbers on a measurement scheme? We can count ten horses or ten cows, or we could just attend to the number ten. Perhaps time is to change as numbers are to counting. Time would not count one second or one minute; time would just count 1-2-3. But Plotinus wonders how numbers measure time. If ten minutes go past, what is the time factor 'ten'? Ten is the number of that time, but how long is that? What number tells us how long ten minutes is? Of course, ten minutes is *ten* minutes. But if time is simple number, 'ten' should be temporally self-explanatory, yet how does 'ten' say how long ten minutes is (how long it feels, how much can happen during it, and how long it *is*)? It does not help to say that 10 minutes is 600 seconds, as if the second number explained the first. For that matter, when motion through time is irregular, as it generally is, what sort of irregular numbers would describe it (s. 9)?

Plotinus concludes that if we look at concrete time first, we cannot explain what time is. We know that time is the moving image of eternity, but we cannot explain time at the level of the image, that is, of the movement, because neither the number of movement nor movement itself explains time as such.

So suppose we try explaining the relation between eternity and its time-image from the other side, eternity (s. 11). Plotinus suggests an odd method. Let us '*ask time* to tell us how it came into being'. Let us ask eternity: what were you doing when you started making images, and why did you need to become? Why were you so restless that you had to start making time, to transfer yourself into time? This seems ridiculous, but the model is that souls desire to create themselves concretely. Animal souls drive them to act in a certain way; humans are driven to reproduce, to form communities, to make things with our hands and ideas. A poet has a mental picture all at once, and then gives it temporal expression. The timeless duty of friendship could take a lifetime or a few minutes to act on. We have a thought, and the urge to make something in time out of it; action temporalizes ideas. In each case, it is as if eternity occurs at two levels: as stable thought-content, and as a project abiding through changes. The second level of eternity, namely life, is more than just living forever; it means having the force that guarantees that whatever one does is part of the same life. Thought is eternal; action is temporal; life is quasi-eternal. When Plotinus suggests that we ask eternity what it was thinking when it made itself into time, it is like asking, how do I act on an idea? Passage into act is 'the spreading out (*diastasis*) of life that involves time' (s. 11).

Time is not just in individual souls, for we see time in the whole of nature. Time is everywhere because soul is everywhere (s. 13). Time is in the soul, but the soul is an eternal house of ideas until it turns timewards. And time appears within the universal soul whenever anything in the universe acts. To discover simultaneous eternity across the world we 'turn back again' (*palin to logo anastrepsai*, s. 12; *palin* 'reverse order'; *ana*: 'turning back'). In typical neo-Platonism, eternity creates us, then we reproduce eternity in our little way, and then we reflect back from what we are doing, and see eternity again acting. The universe is the emanation from One into Many and the return from the Many back to the One. In short, Plotinus connects the abstract and concrete notions of time through the soul's restless activity and backward references.

In short, every act of soul is the simultaneous flowing forth of the eternal idea. The idea is the thing simultaneous with itself; there is no before and after within it, and no delay in carrying it out. The eternal idea, true being, is just what it is, always at the same time as itself, yet ready to pour itself out in successive things that exist not at the same time as it. But because succession is the product of the eternal, at every moment the eternal remains present at the same time as the changes. Time is the moving image of eternity because concrete time acts out eternal life.

Schematically, neo-Platonic conceptions of time anticipate German idealism and phenomenology.[3] For Husserl and Bergson, time is of course explained by individual consciousness, rather than by world soul. But phenomenology too must explain the simultaneity of abstract and concrete time and the simultaneity of simultaneity with succession.

Plotinus's ultimate point is that reality is simultaneous with itself. Eternity is when everything is simultaneous with everything. In modern philosophy, instead of starting with simultaneity and wondering how change is possible, we start with change and wonder how there is any simultaneity left? The difficulty for ancient philosophy is to explain how, in changing things, we see an image, trace, remnant, or after effect related to eternity. The difficulty is that even though the 'is' is different from the 'was' and 'will be', the latter infiltrate the former, and the 'is' starts to disappear in eternal identity. What classical philosophy needs is a middle ground, quasi-eternity, an 'image of…'.

Pasquale Porro[4] documents the thirteenth-century concept of *aevum*, which aims to reconcile the eternal being of God with the things that represent it in perpetuity: eternality outside of time nevertheless expressed in time. It is the time of the angels. Angels are not really in time, but unlike beings trapped in eternity, they can see the world pass by and perhaps move through it. For pseudo-Dionysus, God is pre-eternal (*pro aionon*), whereas angels are synth-eternal (or co-eternal, *sunaidia*) with created eternity (132). Giles of Rome defines *aevum* as *aeternitas participata* (145). Scotus thinks that even finite (sublunar) substances are beyond motion and rest, and so cannot be measured by time except accidentally, that is, by *aevum* (148). For Henry of Ghent, angels change through time (because they perform mental acts), so whereas bodily change is continuous and vague, angelic change occurs in 'discrete time' (*tempus discretum*). Between discrete acts, there is either no interval, or a special intermediate interval (*mora*), depending on whom we ask (156). Ockham, naturally, thinks *aevum* is fiction (149). To be sure, speculating on angels is not a promising way to explain anything. But the theory of time needs a synthesis of abstract and concrete time. It is not enough to say that events are sometimes simultaneous and sometimes successive; something must mediate them.

Without the middle ground of angel-time, how could events in time be visible to an eternal being like God? Standard theology posits that God creates time, but is not *in* time. The simple view is that God sees all of time laid out at once. But if we accept that God can see before and after, because time for Him is all laid out at once, how can God see what is happening in the present? (Conversely, if for God, everything is present, how does God see past and future?) Does God know everything about everything, except what time it is? God may know what all the times are, and what happens at 12:06 on July 31, 2011, but it does not seem that God can know that it is 12:06 *now*.

The conclusion to draw from Plato and Plotinus is that there must be some simultaneity across the drive for change, but that this simultaneity holds across one field where everything is simultaneous (eternity) and another where not everything is simultaneous (time). The symptoms of this non-simultaneous simultaneity take the form of wandering planets or time-souls, but the principle of staggered simultaneity is inevitable for any theory that lays timing measure over timed changes. Changes have to be regulated so that they can measure themselves; the measure produced by change must have been shaping change in advance. The universality at work from the beginning has its origin delayed until after it is used; the delayed effect is in use simultaneous with its pre-condition. Ancient philosophy sometimes deploys teleology to explain how an invariant formal cause gives rise to a succession of efficient causes. But teleology is also a difficult category. The problem of staggered simultaneity is not explained away by, but is articulated in reverse by, the delayed fulfillment of teleology.

(b) Delay as Deferred Eternity: Plato and Origen

Teleology means that the conclusion of an event is already contained within the processes leading towards it. The resolution of a delay is already present in the delay, and conversely, resolution is nothing but the working through of the delay. The telos is the immanent containment of resolution within the delayed phase. Conclusions need not be predetermined (indeed, pre-formation is the opposite of teleological progress). But in teleology, becomings are always working themselves through while they are happening. In one sense, teleology is the science of delay. I am interested in teleology less as fulfillment, and more as the labyrinth of an end that is both never-yet, and also simultaneously a cause all along, the simultaneity of a different time or a time of difference.

What we might call means-centred teleology values the process, resolving its delayed result without ending the delay. In contrast, end-centred teleology will be bothered by disharmonies of speed and deferred satisfaction. By the time of Renaissance literature, delay, or 'dilation' (*dilatio*), becomes a central concern.[5] Thomas Carlson sees 'satisfaction repeatedly promised but ever postponed' as a link between theological promise and consumer culture.[6] Jean Ladrière takes the optimistic view that the paradoxical presence of the *eschaton* makes reason and what it hopes for 'contemporaneous'.[7] But Plato takes a harsh view (*Philebus*[8] 30d–36c) of pleasure-teleology: pleasure is the realm of delay. There is no good pleasure to wait for that would make up for the bad pleasures that get in its way. Pleasure delays a person from getting to reason.

In reason, we have a thought and keep it; we have always already had the thought we are in the process of searching for. There is no delay in the sphere of reason. True, there is delay in retrieving reason now that we have fallen into our lives, having been born and forgotten everything. But delay in reminiscence does not prevent us from possessing all reason simultaneously.

In contrast, the realm of pleasure is temporality. On the surface at least, there is no sense in which we have always already enjoyed the pleasure we are searching for. The tragedy is that once we are in the realm of delay, everything is infinitely delayed. It is not just that each pleasure is temporarily delayed until it is made good, at which point another pleasure is delayed until it is made good. Every pleasure is delayed forever. We get itchy and we scratch, and it hurts, and then it gets itchy again and we scratch, and then it hurts, and then it gets itchy again (46d). There is no end in sight, ever. Pleasure therefore has a twisted kind of eternity. Its delay is not temporary, or temporal, or successive; but neither is it always already accomplished, or satisfied simultaneous with the process of searching for it. Pleasure is eternity delayed *sub specie aeternitatis.*

Plato does not consider the possibility that every moment should be as fully pleasurable as the last and next. He might be right that even constant pleasure would defer fulfillment, but why should that contradict pleasure? Plato's response would have to be that constant pleasure is as much a kind of pain as the lack of pleasure, but this would require a different line of thought, less based on insatiability than on a moral argument about refusing to grow up.

Where is the line between pleasure and reason? Is dissatisfaction a quasi-rational recognition of the relation between being and becoming? Does the feeling of 'bittersweet' intimate reason? Does proto-reason pre-exist pleasure's delay of reason?

Plato sharply distinguishes the lives of pleasure and reason, body and soul, eternity and delay:

> You know that for one who has chosen the life of intelligence there is nothing to prevent him from living in this fashion, a life you mean of neither pleasure nor pain? (33b)

The soul's work has no room for the body. But the twists in the text show the paradox of pleasure in relation to reason.

> Pleasure is itself unlimited, and belongs to the kind that does not and never will contain within itself and derive from itself either beginning, or middle, or end (31b).

In Greek philosophy, of course, limit is good. A perfect circle is circumscribed; it is what it is. Something unlimited is vague, ambiguous, trying too hard to be what it is not, all over the place. Pleasure is unlimited in this way, by nature intermediary, with no beginning or end. Yet it does not even have a middle. Since we cannot define the highest goal of a pleasure, we cannot measure when we are half way there. Relativity is meaningless to pleasure. For a true pleasure seeker, there is no such thing as a nearly good enough pleasure.

This is the first twist. Delay in achieving pleasure implies that we cannot even get close. Delayed for a second, it is delayed forever. As soon as there is any distance to measure, it is measureless. So although we said that time pertains to pleasure and not to reason, it turns out that time does not really pertain to pleasure either. Delay is thus not strictly a temporal category; delay rather prevents the temporal realm from occupying temporal positions.

The second twist concerns what it is we want when we want pleasure.

> I maintain that when we find a disturbance of the harmony of a living creature, that is the time at which its natural condition is disturbed, and distress therewith occurs... Conversely, when the harmony is being restored, and a return is made to its natural condition, we may say that pleasure occurs (31d).

Plato begins with the truism that there is a natural harmony in a living creature. When it gets disturbed, there is pain; when the body returns to it, there is pleasure. Plato does not say that the original state, the harmony we are born with, or even reason itself, is pleasurable. The *return* to it is pleasurable (32a). There is no pleasure in harmony; there is pleasure in a return to harmony. There are two stages in a good life: before the return and after.

Plato is likely speaking ironically. We do not take literally the *Symposium* myth, according to which humans were originally built properly, with four arms and legs, a face in front and a face in back, a double person so powerful and happy that the gods felt impelled to cut us in half, which made us so unhappy that we now perform almost any erotic act to restore that double body. Does Plato think that there was an original time before pain? Probably not. Having a body means being already broken. Falling in love to return to the double state is futile. Returning may be pleasurable, but *being* in a state of homeostatic pleasure is a myth.

So the first twist was that since pleasure has no end, and no middle, it is measureless and atemporal. The second twist is that pleasure never actually moves from one state to another; it returns to an origin that never was. Pleasure is put off into an impossible future. Pleasure is delayed not just because we cannot reach it; it is delayed because its goal is not really in the future. That is the structure of delay: the goal we hope for in the future is actually a past, and a false past at that.

Can a return to the past properly be called a delay? In one sense, Plato's twist is that once something is delayed, it is delayed so infinitely that it is no longer really a delay. The delay loses its force once we realize that we can never make it up, that we are drifting, pulled in all kinds of directions at once, with 'no direction', as Plato describes the dissolute. Pleasure is the disordered realm of becoming in time. Unless time is attached to the eternal truths of the forms, it becomes random, flipping and flopping, and time might as well be going backwards as forwards (as in the *Statesman* myth). In short, the first appearance of pleasure is that there is a particular delay; the second is that we are getting nowhere, so that there is not really delay, but diversion. Pleasure causes delay, and then it destroys delay.

But there is a third twist. When the soul 'anticipates' pleasure, it experiences a second order pleasure, namely confidence in the expectation of pleasure, or else a second order distress as confidence in the expectation of pain (32c). On the one hand, it looks like pleasure is delayed yet again, because before we get to any pleasure, we have to live through its anticipation. On the other hand, the pleasure of anticipation, which we have on the spot, without delay, even if the physical pleasure is delayed, is the greater pleasure. The anticipation of a good meal is better than the good meal, Plato says (though this is not my experience). The good meal makes us bloated, whereas the anticipation, the smell before the taste, is the ultimate pleasurable moment. The twist is that the most pleasurable time is the time when we do not literally get at the pleasure. It does not even matter which pleasure we anticipate; true pleasure is in the anticipatory temporality, in the delay itself.

Indeed, in a fourth twist, anticipation is a faculty not of body but soul, where there is no pleasure, no delay, and no time. So the most pleasurable moment of eating is the moment that anticipates getting out of the whole realm of need and pleasure. That is why it is so pleasurable to anticipate pleasure: for in anticipating, we start to realize that true desire is not for pleasure. We may still think we are getting ready to enjoy taste, but in fact we are getting ready to enjoy reason.

Plato distinguishes this 'intermediate' state between pleasure and reason, from the impossible middling degrees of pleasure (35e). Anticipation is not getting half way forwards towards achieving pleasure, which is not possible, but half way backwards into reason, which is. 'Pleasure of the second kind, which belongs, as we said, to the soul alone, always involves memory' (33b). As far as the soul's enjoyment is concerned, the true structure of anticipation is memory.

It would seem a tautology that pleasure is pleasurable, but in truth pleasure is unsatisfying, since it is a process of returning. Of course, the doctrine of recollection is about the return to the forms, and forms are satisfying. At the level of body, returning to the never-embodied is unsatisfying; at the level of soul, it is. Plato's dialectic typically starts with sensible pleasure, then finds something immanent with which to rise to the next level of soul. In the *Symposium*, first I love this beautiful person here. Then, since I love this beautiful person, I love all beautiful people. And then, since I love them all, I love what is the same about them. I love the universal more than the people. In the *Philebus*, the dialectic is accomplished through dissatisfaction. I seek pleasure; it causes pain; so I hope for the next pleasure and get more pain; finally, I realize I enjoy the anticipation and recollection of the ideal more than I enjoy any state, therefore I will live in recollection. And then I realize I actually did achieve pleasure, I made up for all the infinite delays, by realizing I could bypass the whole bodily realm of delay. Now I live in constant soul pleasure.

The intelligent life bypasses the temporal life by bypassing pleasure. The argument for reason works on this two-sided temporal move: the delay that keeps getting pushed off into the future, and then the restoration of the ideal past that, fortunately, finds no actual time in the past. The person never finds pleasure unless she idealizes the past, so she finds the truth of pleasure in idealization.

Plato distinguishes memory from recollection in order to distinguish the chronological delay that is bypassed, from the ontological delay that takes us back to rational pleasure. Memory is the 'preservation of sensation' (34a); it keeps us in the present of experience, where we retain all our unfulfilled desires. Recollection occurs when something has left consciousness and returns. Memory stores experiences in our bodies; recollection accesses experiences that have left our bodies (34c). Recollection bypasses the beginning-middle-end schema. Because it has no desire, it can rest at the beginning, stay in the middle, or get to the end, of a given experience.

Recollection has a procedure for going back: the repetition of 'becoming empty'. A person who gets thirsty (emptied of drink) for the first time will not 'apprehend replenishment', whereas the soul of a person who has been through full-empty-full-empty cycles opposes its empty body with thoughts of re-plenitude (34e–35c). Just as Heidegger shifts the ground from chronic boredom to fateful vision, Plato shifts from chronically delayed gratification to a different kind of delay. Chronic emptying makes the soul *think of* drink. Ontological emptying makes the soul think; it spills the body out of the soul. For the body, thinking is 'twofold pain' (36a): emptiness paired with hopelessness. But for the soul, it is the fifth twist on delay, a twist called 'simultaneity'.

> When hoping for replenishment we feel pleasure through what we remember, though nevertheless we feel pain simultaneously because of the emptying process going on at the times in question (36b).

In the cycle of emptying, pleasure and pain do not alternate successively, but coincide simultaneously. The body does not recognize that alternation is cyclical identity, but reason does. This makes reason the master of simultaneity rather than the victim of delay. But the sixth twist is that this makes reason—and simultaneity—into an even more substantial delay. For it makes sensory emptiness into the cause and substance of concepts. Soul does not simply exist without body; it is the struggle to empty body out of itself. Reason is not just an alternative to sensation; it is the construction of theories that extract themselves from common experience. Happiness is not just life without desire; it is life on the other side of time. If soul, reason, and the good life pass through to the other side of time, they do so at the moment of transition, that is, without delay. But time as such is an ontological delay of the presence of reason. If we had to wait a certain amount of time for reason to show itself in experience, reason would be just another pleasurable sensation, the desire for which would never be satisfied. Reason must rather present itself eternally. But it can only be eternal on the other side of time, time perpetually emptying itself of time. 'Time is the moving image of eternity': not as an unwinding spring, but as simultaneity running simultaneously with the fact that each and every process constitutes its delay. The rational cycles of empty time are independent of the stops and starts, the shudders and relaxations, the accelerations and decelerations, of filled time.

In sum, the paradox of the concept of delay in ancient philosophy is that it draws out time, but cancels time in favour of eternal simultaneity; yet the eternal enters into time just at the moment of maximum delay. The idea of eternity on the other side of delay seems pre-modern. Modern philosophy will want to synchronize simultaneities in physics, perception, and desire, without resorting to eternity. At some level, ancient philosophy seems right to say that the measure of time must be an eternal abstraction, that while events may be delayed, time itself is neither the delay, nor the return, nor the catching up, of anything. In Plato's terms, reason eschews the time pleasure takes, yet reason is the principle behind the backward teleology of pleasure's return. Reason would be the pure law of time, without taking place in time. In Aristotle's terms, time measures speed but does not move quickly or slowly, or at any speed.

In *Physics*[9] Book IV, Chapters 10–14, Aristotle says: 'Change is always faster or slower, whereas time is not' (218b, 220a–b, 223a). In fast change, more movements take place in a certain amount of time; in slow change, fewer. Movement is change in things, and time counts that movement (218b). The speed of motion is defined by time. But time is not defined by speed. If time had a speed, time would be measured by time, which is absurd (218b). If time could be fast, more time could take place in the same amount of time, which is absurd.

But there are provisos to this doctrine even in Aristotle.

Time is not movement or change, Aristotle says, but neither is it separable from movement (218b, 220b). Time 'belongs to' movement (219a). It is not just that time can only be perceived when there is movement (219a). It is that time is not merely an abstract measure by which we count movements; time is what is count*ed* (219b). But if time is counted, and not just the rule for counting, then time is in some sense concrete after all; time is the substance and not just the form of movement.

Indeed, much of Aristotle's chapter is a topology of time. Movement is a continuous magnitude: between any two points, there are intermediate points (219a). Time is in the distance between two extremes, the middle range between a beginning point and an endpoint. Indeed, a now marks the beginning of some movements, the middle of others, and the end of still others. It follows that there will always be time, that time will 'never fail', since every now-point is the beginning of something (222b). Every now exists relative to others before and after (219a). Each point in time is a link (222a), dividing the time-line into segments, stretching time across virtual intermediaries (220a).

Time is counted by relational numbers, not by simple cardinals. Aristotle likes to say that the smallest number is two (220a). The elementary number is a two-place co-ordinate set, marking the two limits of a vector. The numbers needed for measuring time might be less like 1, 2, 3... or 1st, 2nd 3rd..., and more like 2/4, 3/4, 9/8, that is, more like time-signatures in music.[10] Since time measures where a change begins and terminates, we might even call time the measure of delay. But for Aristotle, events are not unfulfilled, as in Plato. On the contrary, events are fulfilled in the topology of time, potentially infinite delays are given as continuous actualities.

The difficulty is that if time is counted by time-relations that are already counted in a different way, then Aristotle's presumption that time cannot be defined by time (218b) is not entirely true. Speed might in some sense be attributable to time after all.

To put it differently, time-measure treats time as a multiplicity, and only secondarily as a now (220a). Aristotle does not want this to undermine the unity of time (223b). Of course, many movements are simultaneous within each now, and each now functions differently within different series (beginning one series, ending another). Indeed, this is how movements can be fast or slow. But time itself, he insists, is the 'substratum' for all temporal series. Time is the substance, and each now is an attribute of time (219b). Citing the Stoics, Aristotle says that 'the "now" corresponds to the body that is carried along, as time corresponds to the motion' (219b). It is not just that time is what is counted in movement, and that movements are temporal because they can be assigned numbers on a time-line; it is more that time is a subterranean stream along which bodies are carried. Nows are like temporal bodies that inch their way along prime temporal matter. Aristotle's goal is not just to dissolve a reified ontology of time and to put in its place a formal epistemology; his goal is to give scientific descriptions of concrete temporal categories (222a–b). To cite just two examples, he defines the temporal 'near and far' by measuring distances indexed to nows that might be past or future as well as present. And he defines 'sudden' by measuring temporal minima, imperceptible temporal intermediaries, such that a later event appears to begin with a jolt, without delay.

Aristotle is trying to say two things. First, he says that time is abstract: 'being *in* time' does not imply that time is a substance with which movements coexist but only means that movements need not always occur (221a). Second, he says that time belongs to movement: things are 'affected by time', in that time is 'the cause of decay, since it is the number of change, and change removes what is' (221b). But then a few pages later, he goes back to the first, emphasizing that time is abstracted from movement: though we speak of 'a thing's being destroyed by time, nevertheless time does not work even this change; even this sort of change takes place incidentally in time' (222b). Aristotle's subtlety is carefully worded: 'Time is not number with which we count, but the number of things which are counted' (220b). Time is not number. Time is not things in movement. Time is the 'number *of*'. Time is number intentionality.

In short, Aristotle's separation of time and movement, and his association of time with abstract number rather than with concrete change, imply that time has no determinate speed. To some extent, common sense has to agree. Yet the fact that Aristotle defines the peculiar temporal numbering system in terms of links and limits, nearness and jolts, and the of-ness of time implies that speed is at least co-extensive with time and, indeed, that a multiplicity of relative speeds is co-extensive with the relational numbers that measure time.

Plato tries to put concrete eternity on the other side of concrete time. Aristotle tries to put abstract time on the other side of concrete movement. Plato puts delay into time whereas Aristotle takes it out. But in both cases, it is differential speed that makes the other side of time perceivable. What would it look like to put delay and speed differential into eternity itself, into the very definition of time?

Christianity's personalization of Being offers a case study, by temporalizing the distance between time and eternity. Giorgio Agamben's *The Time That Remains*[11] analyzes the temporal situation described by Saint Paul: the end of time has been revealed but has not yet occurred, so the Christian lives in the ambiguous zone of 'the time remaining'. In general, neo-Platonism treats the middle period between temporal emanation and the return to eternity as a delay zone. Paul's epistles posit a 'time remaining' between the resurrection of the messiah and the end of time. '"In the time of now (*en to nun kairo,* the technical expression for messianic time), a remainder is produced by the election of grace" (*Romans* 11:5)' (95). At one level, Paul is talking about the 'remainder' of Jews who will be saved—it will neither be the whole of the Jewish people nor any specific minority; the remainder is the division that time produces within a people (96–8). At another level, Paul is talking about the scission of time. Pre-messianic time is normal succession, part by part. At the end of time, God is 'all in all', and time will have no parts. Messianic time, the 'time remaining', takes place between these two, neither a part nor the whole of time. Messianic time is the caesura of time at the moment of choice (98–9). In a sense, all lived time has this character of being between part and whole, which means messianic time is the only time there is (100), the only time remaining to us, the time that makes it possible to speak of any temporal series having a result, the time that lets us do what is required for time to come to an end (119), 'the time that remains between time and its end' (111).

Messianic time is thus not exclusively about the prophetic future but also about the present now (110). A simplistic Christian history would divide the ages of time into (a) the prosaic chronology of world history before the messiah is resurrected; (b) the 'contracted' time that we live in now after the resurrection while we wait for the world to end, while the end is 'delayed' either by Rome or by God's will (184–6); then finally (c) the end of time when the full *parousia* of the messiah explodes into eternity (112–3). But Agamben argues that tenses should be understood less in terms of past, present, and future, and more in terms of what the linguist Gustave Guillaume calls 'operative time' (116–7). Time is always at the moment of being constructed in language, and then, after a 'delay', in subjectivity (118). The operation of time, the dynamic that lets time pass, is not another time before or after any successive moment, but time's 'interior' (119). Seizing the moment (*kairos*) to set time in motion reflects our experience of lived time, as well as Paul's description of messianic time (120–1).

There is thus a 'constitutive delay' in the nature of time (123) that does not just point to the future but triples the present's operation. But I want to consider how the delay between time and eternity enters into the period of eternity itself. Teleology does not just defer an endpoint but can also locate deferral inside the endpoint itself, inside the afterlife.

It can seem that the whole point of eternity, virtue, happiness, and reason is to avoid delay: ontologically, ethically, psychologically, and epistemologically, as if an all-knowing God puts the world into a non-delay state by means of omni-temporal omniscience. In one sense, for Christianity, the second coming is delayed, but that delay can put us into pre-existing contact with the eternal. The clue that there is more to time than eternity is found is the life history of sinners, even repentant ones. For if we delay repentance, then we do not get to use the delay of the second coming to become good in advance. Conversely, if sinners delay actualizing their potential for sin for the duration of their lives, the punishment they deserve on account of original sin may be commuted forever.

The basis of penology is that a crime should carry an expectation of punishment; punishment should be virtually simultaneous with the crime. Even if the actual punishment is delayed, or the criminal is never caught or is unjustly pardoned, ontologically speaking, crime *is* deferred punishment. As Derrida shows in the rhetoric behind the death penalty, the condemned man ought to consider himself his own executioner.[12]

But if there are rewards and punishments, why should God wait until we die before punishing us? Perhaps this temporal mercy gives us a chance to back off before passing from intention to action, or to make restitution after the fact, or to go to the police and explain ourselves before the punishment is decreed.

But Origen shows the paradox. God ought to teach us what we should do before we act; sometimes He delays in teaching us, which means that He will have to teach us what we should have done after we have already acted but before the punishment begins. And if that belated teaching works, He may not have to punish us after all, after we no longer deserve it, in spite of our having previously deserved it. Delayed punishment entails an odd account of moral agency. Origen's *On First Principles*, Book 3, Chapter 1[13], asks what made Pharaoh persecute the Jews? Which came first: Pharaoh's bad intent or God giving Pharaoh the character that made him have a bad intent? As Origen puts it, what does it mean when the Bible says that 'God hardened Pharaoh's heart' (*Exodus* 4:21 and elsewhere)?

The punishment of Pharaoh could have happened sooner—which would have prevented Pharaoh from doing wrong, and therefore would have reduced his punishment later—but that would have been too early, it would not have made enough of a difference. Indeed, when people are punished or rewarded earlier than they should be, they sense it is not fitting. Origen speaks of a farmer who takes time planting seeds because he does not want the crop to grow too fast. Similarly, rather than sealing a wound right away, a physician may leave it open, and let the body heal naturally in its own time, 'to defer its immediate healing with a view to ensuring better and sounder health, since physicians know it is preferable to cause delay in wounds that fester, and to allow the malignant humour to flow for a while, rather than to hasten to a superficial cure' (181–2 and 191). It takes longer, but using the body's own power heals better. The faster way is slower. Delayed time is the fastest there is.[14] Delay is the right time and the alternative is premature.

But why did God delay punishing Pharaoh? If God had punished Pharaoh before he persecuted the Jews, all that suffering would have been avoided. The strategy of theodicy is that God had to allow extra disasters on the planet so that... But it is not only that God allowed Pharaoh's hard heart to go into action without obstacle. Pharaoh was about to soften and allow the Jews to go free, but instead, God hardened Pharaoh's heart so that he would persecute the Jews further. Is God Trotskyist, reasoning that if Pharaoh refuses to let them go, then the Jews will revolt, which will lead them to Israel?

Origen exhorts us not to think that the end of the story occurs when the narrative of action ends. It is true that God hardened Pharaoh's heart, made him a worse person, so that Pharaoh would deserve to be killed. But just because Pharaoh dies does not mean that God has finished with him. In Origen's account, the afterlife is a lecture room for the reeducation of souls. Pharaoh will be retrained after dying, and eventually will be rewarded for that. In the very final analysis, it is not that bad a deal for Pharaoh, or for the Jews who died.

In the third century, it was not obvious whether the Christian God was the same God as the Jewish God, or that the New Testament scriptures belonged in the same book as the Old. Origen's position became the standard: the prophesy of the Old Testament is after a delay revealed by the New (190). The Old Testament is finally fulfilled; it was the truth, but in a form whose self-evidence was delayed until later.

This is related to the question of when Pharaoh turns bad and when he deserves to be punished. If God created him as a certain kind of soul, then he would not have original free responsibility for his decisions. Everyone knows that in many ways God does not determine a person's decision-making; He gives him a soul and lets him decide what to do. But Origen is willing to take the hard case. God did not merely create Pharaoh with free will; God took a person and made his heart hard, and now the person is blamed.

> We are going to see what the prophets declare about those who, by living a good life at the first have deserved to receive very many proofs of God's goodness, but afterwards, as men do, have fallen to sin. The prophet, making himself one with them says 'Why, Lord, didst though make us to err from the way? And why didst thou harden our heart that we should not fear thy name?' (177–8)

They were willing to fear God's name and to do His bidding, so why did He harden their hearts so that they would not? When they were good at the start, they feared God, so they were willing to be good. The allegation is that if God had kept up the level of fear, they would have stayed good. If God had only 'chastised' them constantly, as one constantly beats a horse so it remembers to go (178), they would have remained good. Why did God stop hurting us? Why does He neglect to punish some good people? God's crime is that he went easy on us, thereby making us deserve worse punishment. This is the essential question. When did we deserve to lose the temporal momentum of our good start? Origen distinguishes three stages in the temporality of Pharaoh's first wrong decision (178–9). First, God knew in advance what Pharaoh was going to decide. From God's point of view, there is no question of Pharaoh starting off good and then having his heart hardened. God is outside of time, He sees what we do at all given times, He sees events on McTaggart's B-series, with dates attached, but not one at a time. Second, then, Pharaoh's choice falls under what Origen calls the 'steadfastness' of things. It is only we finite minds living in time who see things in the third way, namely, as first coming into existence, then turning bad, then deserving punishment.

The problems, in short, arise only from our own temporal point of view. This is not so much anti-temporalism as what we could call pre-positionism. We do not know that our good beginnings come from God unless until we 'first come to the point of desiring to obtain a benefit' (179–80). God is always already giving us a chance to act well, but we do not know this. We do not know God's instructions unless we desire to receive them. God has given us his Old Testament, but we only get it when we read it as a New Testament.

> Which will only happen thus, when a man has previously learned to know himself, and to feel what is lacking in himself, and to understand from whom he should and can seek the supply of what is lacking. (180)

In short, God gives us some knowledge about how to live. We can only accept this knowledge if we desire it, and we can only desire it if we *previously* learned to know ourselves. The circularity is a desire-version of Plato's reminiscence. We can accept God's instruction only if we previously accepted God's instruction. This is not just Origen's confusion: it is required in order to explain moral agency in relation to God's creation. Moral agency is consistent with God's having created us in a certain way, only if that kind of creation is what we wanted. And we wanted it because God created us that way, but that too is what we wanted. If we do not desire God's gift at the start, then later we will become arrogant, which will have made us not desire God's gift. Revelation is thus a return to childhood beginnings. It is hidden except for

> those who, after becoming infants and babes, or in other words returning to the simplicity of babes, then make progress, and when they reach perfection remember that they have obtained their blessedness not so much by their own virtues as by the grace and mercy of God. (181)

In brief, return brings progress, which brings memory.

This would seem too paradoxical and unhelpful for moral agency, if it were not for the fact that the soul is immortal, and has plenty of time to cycle through these returns before being judged. 'Being an immortal and eternal thing it is not, even though it be not quickly healed, thereby shut out from salvation, which is only delayed until more convenient times'

(181). Delay is the cure for succession; delay returns us to eternity. Slow reception, longsuffering patience: these practices are not only the temporal soul's way of imitating eternal stability, they are also God's method of introducing justice into eternity.

If we do not take temporal recycling of choice into account, Origen's pronouncements on guilt seem merely contradictory. To the question, 'why do certain seeds fall upon rocky ground' and others 'spring up too rapidly', or why do people not make the right choices until they do, Origen alternately appeals to the soul 'becoming aware of its fault', and to 'God alone determining the time for its healing' (184–5). How can we choose to be good if God alone can determine the time when we become good? If it is 'not in our power to put away… the wickedness by which a man is hardened' (186), how can it even be in our power to ask God to help us? The only way out is to posit a backwards referring decision, like Augustine retroactively blaming his childhood self for selfishness that no child could help, or like a bystander blaming himself for not having gotten involved in something he was too stunned to react to, or like a person letting someone else be the person they had always been searching for. The other, the future, or the insight, becomes the source of one's prior agency.

> If he, who had formerly hardened himself in ignorance, entrusts himself, as we have said, with full purpose of mind to a master and promises to obey him in everything, the master, on seeing clearly his purpose and determination, will on his part undertake to take away from him his lack of education. (187)

Obedience to the master and 'full purpose of mind' imply that the mind is a promise and not a content, and that one acts not on the basis of education already received, but in order to become educated in what one should have been doing.

If goodness does not 'delay the healing and defer the remedy', the error will fall into 'repetition' as soon as the medicine is withdrawn (191). As long as the afterlife is an indefinite proliferation of education, there will be no point at which a sin is ready to be punished. As long as the softening and hardening of hearts is still in motion, cycling through self-reflection, evil intent is delayed. As long as delay keeps time from hardening, a hardened heart is still supple. Delay allows for the mutual causality of soul and God, of temporal and eternal causality. Or better, delay is the symptom of a cure working for eternity rather than for a certain time. Circularity avoids repetition.

While responsibility absolves ultimate guilt (consistent with Origen's heretical thesis that there are an infinite number of created worlds before and after this one, in which the same souls make their choices again and again, before time finally comes to an end and God becomes all in all), it wildly expands the room for regret.

Aristotle, with his usual common sense limits, says that we can only regret acts when we could have done otherwise. A person living today, Aristotle says, cannot regret failing to have fought in the Trojan War, an event that took place centuries before he was born. After all, one cannot change the past. There is no point wishing that one were 10 years younger, so how can there be a point to wishing that one had not behaved as one did in the past? But perhaps there is a sense in which every possible action is already in some way regrettable, and in particular, that choices in the past remain live options, that memory reactivates the past by will—not just as a series of facts, but also as a project of self-interpretation. It may be essential to memory that regret apply in different ways (a) to every action, (b) to no actions, and (c) to certain definite actions.

In the *Nicomachean Ethics*[15], Aristotle says:

> No object of choice belongs to the past; no one chooses to have sacked Troy. For deliberation does not refer to the past but only to the future and to what is possible; and it is not possible that what is past should not have happened (Book 6, 1139b).

I emphasize three points.

First, an act is voluntary when the initiative or source of motion comes from within and involuntary, or constrained, when it originates from without (1110a). Aristotle suggests that while we should forgive people when they act wrongly at a time when they are 'under a greater strain than human nature can endure' (1110a), the pleasures and pains accompanying actions do not alone make them involuntary. Otherwise, all actions would be involuntary, which is absurd (1110b). Aristotle's position assumes that an agent knows what is within her power and what strains she can endure. Self-knowledge shapes the criteria for involuntariness. One is not responsible for one's action if one is truly ignorant of such facts as who the agent is, what objects are affected, what means are required, and so on. Aristotle immediately notes that the first in his list, the ignorance of the agent, is impossible, 'for how could a man not know his own identity?' (1111a).

Second, regret is a clue to the voluntary. For Aristotle, an action committed in ignorance is involuntary only 'if it also brings regret in its train' (1111a).

If the agent later discovers the missing knowledge, and does not then regret the deed, then it was not ignorance that had shaped her choice, but a lack of principle or character. Regret is the sign that a past action was involuntary. It is only when a gap is opened between the agent and her past action that an agent can see the border between what is and what is not in her power.

Third, while the test of what is involuntary may only be conducted on the past, the choice (*proairesis*) is future-directed. Now, since we can choose and deliberate only about what is in our power, and since we do not have the distant future within our power, Aristotle says that we can choose and deliberate only about the means of our actions, not about the ends (1112b). This seems odd, but the idea is that ethical thinking is like starting with a desired endpoint and thinking back along a chain of possible causes until one gets to something one can oneself do now. 'Every man stops enquiring how he is to act when he has traced the initiative of action back (*anagagein*) to himself' (1113a). Choice is futural, but deliberation directs thought backwards from the future into the future's possible pasts, namely, the present. If the end is never achieved, then when we get to the future, the memory of our actions will branch backwards into a causal chain ending in an action we will then regret.

In short, since the voluntary is about deliberation, and deliberation is about retracing steps, the past is what we take responsibility for. Like the already deserved but too early hardening of Pharaoh's heart, choice is about setting the grounds for regret for actions that will be too late to control.

In fact, there are difficulties in Aristotle's account. It is not clear that we can know the limits of what is within our power. Nor is it clear what pains we can endure and still remain free. Nor is it even clear what it means to know one's own identity. When Aristotle says that to deliberate, an agent has to 'trace the action back' to himself, it is not clear to what exactly he is tracing back. Each choice requires that one be able to interpret the scope of one's own possibilities and powers. But one's own powers are not determined or known wholly from within. For that matter, one's interpretations of one's powers depends on how one interprets the choices of others. And interpretation does not occur in a flash but emerges over a lifetime of decision-making. The scope of the agent's power is not settled before specific decisions are made but comes to light in the history of those decisions. In short, to trace an action back to oneself is to trace a history of one's own self-interpretations as well as the self-interpretations of others.

So might someone today regret having not fought in the Trojan War? Certainly, one cannot change the past. But the extent of the power of one's will is not something one knows immediately; it is known in the successes and failures of constructing a possible means-end chain. One will never succeed in acting towards fighting in the Trojan War, but if that war has significance for one's self-interpretations, one might expect on occasion to trace oneself back to the chains of events originating in that war. It will engage one's memory in such a way that one gages one's powers in relation to what one imagines one might have been able to do in that context. It is not only a matter of comparing one's will with that of others in the past. It is a matter of tracing one's powers back to their origins, and the knowledge of our powers lies both in what we have done and in what we have not, both in what we are satisfied with and what we regret, both in what we have done ourselves and in what others have done in what might have been our place. Contrary to Aristotle, it might be quite important for our future choices that we regret not having fought in the Trojan War. The sense in which we might regret actions about which we could not have done otherwise is the sense in which regret for the past is delayed until choices about the future have been traced back to the present.

There is a proverb relating to Go, the Japanese board game: 'Always make your opponent regret his last move'. How is this possible? If your move is good, can your opponent not have avoided it by moving elsewhere in his previous move? Is there not some move your opponent could have made that would not be regrettable? If not, what would he regret his last move in favour of? The point of the proverb is that every strategy must be flexible, and every move must have multiple meanings. The object in Go is to surround more space (more 'territory') with your stones than your opponent is surrounding with his. The player balances gains of small but secure territory with expansive but loose influence. With each move, the balance within each player's position shifts. If you can lure your opponent out of balance, the game shifts in your favour. Since each move potentially can be used either for territory or for power, you have to have two plans for each move. In case your opponent interprets your previous move as a claim to territory and moves to block its territorial value, you have to have prepared a plan to follow through in your move's other, influence building, meaning. In a good game, both players are constantly reversing the meanings of each other's moves. In a good game, both players regret all their previous moves, even when each was the most powerful possible.

With the idea that consciousness can retroactively take responsibility for a convergence that existed at an earlier time and that simultaneity is built on the freedom to move through time, we move to the modernist problematic of Kant and Hegel.

Chapter 4

Simultaneity and Delay in the Era of Time: Late Modern Issues

Theme of the chapter: The test of whether two objects are simultaneous is whether a perceiver can freely return from one to the other. Delay breaks free from the circle of simultaneity.

We have seen that staggered series have to be synchronized in order for events to be simultaneous, and those synchronizations lead back to simultaneities in the past and future. Many cases of simultaneity involve events in the same experiential field. But we can also experience the simultaneity of a presently experienced object (a room in a house) with something not presently experienced (another room in the house), but possibly experienced at another time. Kant defines simultaneity through the possibility of going back to look at something seen before.

(a) Simultaneity in Kant: The Schema for Returning to the Same Place: Simultaneity as Reversible Succession

If I see one thing and then a second, then return to the thing I saw before, and see that it is still there, I know that the first thing has been existing simultaneously with the thing I was seeing more recently. Two objects are simultaneous at time-2 if (a) one was experienced at time-1, (b) the other is experienced at time-2, and (c) it is possible to experience the first again at time-3.

Kant's proofs for the categories are contained in the 'Schematism' chapters of the *Critique of Pure Reason*[1], which also explain how the categories are applied in time, and in a sense construct time. The schematic proof of the category of causality involves the construction of succession; the proof for the category of community, that is, reciprocal causality, involves the construction of simultaneity. Kant's examples of simultaneity and succession respectively involve the rooms in a house and a boat going down a river.

When I walk through a house, I walk into a certain room; then I walk into another room. My experience changes successively with time, but I can see that the house is there all at once. In the case of seeing a boat first upriver and then downriver, the succession in my perceptions is correlated with succession in the object. What operations make us judge that some objects that we see successively also exist successively, while others that we see successively exist simultaneously?

Kant's schematism of simultaneity involves reversibility, return, and delay.

> Things are simultaneous if their perceptions can in empirical intuition succeed one another reciprocally.... I can carry on my perception either first with the moon and thereafter with the earth or vice versa, first with the earth and then with the moon. And because the perceptions of these objects can succeed each other reciprocally, I say the objects exist simultaneously' (A211/B257).

The order in which I choose to perceive the earth and the moon, or the rooms in a house, is indifferent and reversible, and this proves that the rooms exist simultaneously. Conversely, when the states of an object exist at different times, I cannot perceive them in backwards order.

A house is too big to see all at once. At *this* time, I am seeing only this room. So how do I know that the other parts of the house are there simultaneous with what I actually experience now? The answer is that I go back and look again at what I saw earlier. (Presumably, if several of us are there simultaneously, we apprehend the simultaneity simultaneously.[2]) If Room A is there again, that is my proof that A continued to endure all along while I was looking at B—though of course now that I am looking at A, I would have to return to B to know that B has been there while I have been looking at A ... and so on, reversing the order again and again.

To be sure, the time it takes to go back produces a new temporal referent. But that is the point about simultaneity. I cannot actually see the two distant objects at the same time. By the time I go back to the next room to test whether it *was* simultaneous with this one all along, the time when I saw the last one is no longer accessible; yet I legitimately experience the other room as having existed in the duration. The 'interim' is crucial to the evidence for, and the logic of, simultaneity. It is always the next perception, the delayed perception, that is the test of simultaneity. The result of the test refers back through the time between first and last perceptions. The period of time during which only one thing is perceived has to demonstrate the simultaneity of the two things.

As Kant says, 'time cannot itself be perceived'. I cannot experience the other time at the same time I am experiencing this. All I can do is set up a scenario, a synthesis of anticipation, a hypothetical conditional for simultaneity that is apprehended but unperceived. *If I return* and look, and it is still there, then I can say that it *was* there in the interim. Simultaneity is a possibility-sequence that I do not actualize all at the same time.

To be sure, there are technical difficulties. When I look back and do *not* see the thing seen before, does that tell me that while I was looking here, it was not still there? Maybe that thing remained simultaneous with what I was looking at for quite a long time before it disappeared. In fact, certain cases will systematically produce deceptive results using Kant's test of looking back. If I look back and forth from one end of a teeter-totter to the other and I happen to turn my head back and forth in a tempo synchronized with the swing, then each time I look at either end, that end will be up. I might think that the two ends are always simultaneously up. Heuristically, I should need to vary the tempo of back and forth looks to get the best results on a simultaneity test.

Of course, pragmatically, when I am in one room of a house, it is ridiculous to bother confirming that the other rooms are still there. I rely on the a priori possibility of returning. Yet, apriority alone does not guarantee simultaneity, since simultaneity is what Kant calls a 'dynamic' category. The 'mathematical' categories, like unity, pertain to every possible object. The 'dynamic' categories, like cause and simultaneity, need not. Not every pair of things is simultaneous, obviously; we have to test. Indeed, there is some small degree of uncertainty about whether the room beside me is still there at the same time as this one is. Maybe it has burned up in a fire that has not yet reached me. There is no absolute proof of simultaneity, only a temporal structure that makes it *possible* each time to check again.

In most cases when we assert simultaneity, the test has not yet been carried out. This is why the establishment of simultaneity is in each case delayed. Simultaneity is an indirect experience, reflective rather than assertoric, active rather than immediate, unlike the experience of succession. When I walk from room to room, I directly experience the sequence. I am right there as the sequence happens and things change. There is nothing I need to do or control in order to perceive succession, other than to notice it. To experience simultaneity, in contrast, I have to overlay a meta-temporal simultaneity-structure onto the temporal succession-structure that I experience. In order to say, 'the other room is there now too', I have to be able to control where I go next, to control the order of my experience, to verify a temporal property of objects in two places precisely when it does not matter which place I am in at the time I affirm it. When I am active about how I see things in time, I add an arbitrary decision into the time-flow. That it does not matter where in space I am at a given time, and that I do not have to care which room I go into: that is the test of simultaneity.

The subjective procedure (controlling the reversible order of perceptions) tests for the objective simultaneity of the rooms. But we will now see that the subjective test presupposes another, objective test. In this passage, the reason why simultaneous things are indifferent to what order we perceive them in is that things that are simultaneous are outside one another.

> Hence for things existing outside one another simultaneously we require a concept of understanding of the reciprocal succession of their determinations, in order to say that the reciprocal succession of the perceptions has its basis in the object. (A211/B257)

There has to be something in the *object* that is simultaneous, not just something about our subjective capacity to perceive things in arbitrary order. So what is simultaneity in the object?

> The relation of substances wherein one substance contains determinations whose basis is contained in the other substance is the relation of influence. And if this latter thing reciprocally contains the basis of the determinations in the former thing then the relation is that of community or interaction. Therefore the simultaneity of substances in space cannot be cognized in experience except under the presupposition that they interact with one another. (A211/B258)

If two things are simultaneous, then each is in part the cause of the other, or each contains determinations of the other. Earth and moon have reciprocal gravitational pull. The moon is what it is, and where it is, while I am looking at it just because the earth, which I am not looking at, is still there keeping the moon in its orbit. If two things are simultaneous, they must not only be there together in space, they must also have causal influence on each other. When two things occur in succession, the first has causal influence on the second but not vice versa. Simultaneity implies reciprocal causation.

In fact, the idea that causes are simultaneous with their effects raises a problem. Kant's example of a ball causing indentation on a pillow and Aquinas' example of a person causing his pencil to write seem obvious cases of effects that occur simultaneous with their causes. Against such cases, Hume argues for 'the priority of time in the cause before the effect'.[3]

Assume that a cause could 'in the very first moment of its existence, exert its productive quality ... perfectly co-temporary with its effect'. On that assumption, any cause that 'retards its operation for a single moment' would by definition not be a sufficient cause. If a cause is sufficient, its effect should occur without delay. On the surface, it seems obvious that causes should have their effects right away; it seems absurd that after the cause appears there should be a gap in time when nothing happens, and that after a small pause during which nothing relevant happens, the cause should suddenly start acting. Causal influence delayed for no reason is certainly peculiar. But for Hume, immediate causal influence would be even more peculiar. For if sufficient causes are co-temporary with their effects, then all of those effects will be co-temporary with the further effects for which *they* are sufficient causes. All causes would be co-temporary with all subsequent effects ad infinitum. The entire history and future of effects would occur in the same instant, resulting in the 'destruction of that succession of causes' and the 'utter annihilation of time'. This is truly absurd, Hume says. Instead, the existence of time depends after all on there being a 'retard' between a sufficient cause and its effect, a delay between the existence of a cause and its operation, a delay between what is present in a moment of time and what takes place because of what is present at that moment. One might almost say that a delay is one of the causes in the causal series.[4] It is not just that Hume rejects the (traditional) view accepted by Kant, that a cause and its effect can occur simultaneously. He also rejects the smooth continuity of filled time in favour of a jumpy series of delayed reactions.

Perhaps the reason why Kant and Hume disagree about whether cause-effect relations can be simultaneous is that while they both think that causality is a narrative constructed in understanding, and not an immediate experience, Hume assumes that simultaneity is immediately experienced (too immediate to contain causality), whereas Kant thinks that simultaneity too is constructed out of successions, lacunae, decisions, reversals, and counterfactuals. Subjectively, Kant says, it takes time to trace a succession back to simultaneity, and objectively, causal simultaneity is what allows the world to have remained there when we go back to look for it. The twin subjective and objective models allow for a simultaneous world to undergo successive states without annihilating time or collapsing into *totum simul*.

Nevertheless, the relation between Kant's subjective and objective versions of simultaneity creates a problem (just as the respectively subjective and objective A- and B-transcendental deductions make an uneasy conceptual scheme). There is no problem in saying that subjectively, perception is filled up one room at a time into the representation of a whole house; and also saying that objectively, each room contains determinations of, and influences on, the other rooms. But causal simultaneity has transcendental presuppositions different from those of reversible perception. The causal conception of the house includes, for example, the physical dependence of the third floor room on the ceiling of the second floor room. If the ceiling of this room is not collapsing, that is a good hint that the floor of the room above is still intact. The subjective presumption that I *could* reverse the order of my perceptions and look again at the room above is not strong enough evidence that the other room exists simultaneous with this room. To judge that the room above me has not collapsed while I was not looking, I cannot just imagine climbing the stairs, I also need to assume objective causal influence. In short, Kant appeals to two independent sets of assumptions behind judgements of simultaneity.

For that matter, succession creates as many problems as simultaneity. The paradigm of succession begins plausibly: a boat goes down a river in successive stages; after I have seen it downriver, it is impossible to look back and see it upriver again. But by the subjective test alone, how could I know that it is impossible to reverse the order of perceptions? It seems unlikely that the boat would suddenly reappear up there, but it is not as though Kant believes that any empirical fact is necessary, so why is this impossible? Pragmatically, if I did unexpectedly experience the positions of the boat in reverse order, satisfying the subjective test of simultaneity, I would nevertheless resist inferring that the positions of the boat were objectively simultaneous. More likely, I would posit some fast and invisible cause making it float back up. Yet if I am ready to interpret reversible perception as the result of speedy causal production, irreversible casual explanation might replace simultaneity in the case of the house too. Seeing the front hallway of the house a second time could be interpreted not as the simultaneous existence of the rooms, but as the result of a speedy rebuilding of rooms, each of which collapses immediately after it is viewed.

At any rate, Kant offers only these two reasons for thinking that two things are or are not simultaneous. The subjective test is not convincing until the objective criteria are met. And even if we hypothesize mutual causal determinations between two things, we still have to be able to confirm them by going back to look. Subjectivity and objectivity in experience are each delayed pursuant to the other.

Perhaps simultaneity requires a stronger relation between subjectivity and objectivity in space and time. Kant raises two hints, one on space and one on causality.

In terms of space,

> Now suppose that in a manifold of substances, taken as appearances, each of them were completely isolated, that is, that no substance effected influences in another and reciprocally received influences from it, I say in that case their simultaneity would not be an object of a possible perception, and that the existence of one substance could not by any path of empirical synthesis lead to the existence of another. (A212/B258–9).

The possibility of going back and seeing a room again means that there is always a 'path' to get back, which a subject can negotiate through objective space. This has four implications for space.

First, space has perspectives and directions in it, mapped pathways and relative positions. The order for viewing simultaneous objects must be indifferent, but once I decide to look back, I must be able to *get* there. Second, space must be full, with no unleapable ditches or vacuums. It is not just that some one room is simultaneous with the room I am in — all of them must be. The fact that my choice of direction is arbitrary means that all of space is universally simultaneous. The alternative would be weird, as if I were to come to a place that was not there now but was there at a different time. Third, space has to be infinite in extent. 'No matter how remote' something is, it has to be simultaneously there. Fourth, *I* personally have to be able to move through it.[5]

Just as there must be a 'spatial community' (*communio spatii*), there must be a community of causal interdependence (*commercium*) (A212–3/B259). All the world's causal influences have to interact at the same time, in a sense causing each other's simultaneity.[6]

> We can easily tell by our experiences: that only the continuous influences in all positions of space can lead our sense from one object to another; that the light playing between our eye and the celestial bodies can bring about an indirect community between us and them and can thereby prove their simultaneity; that we cannot empirically change place (and perceive this change) unless matter everywhere makes possible the perception of our position; and that only by means of matter's reciprocal influence can matter establish its simultaneity and thereby establish (although only indirectly) the coexistence of objects, down to the most remote ones. Without community every perception would be severed from any other; the chain of empirical presentations—i.e. experience—would begin anew with each new object, and the previous chain could not in the least cohere with it or stand to it in a time relation (A213–4/B260–1).

This has three implications. First, spatial positions are coordinated by the simultaneity of causes and effects. Second, change of place depends on position, and hence on the simultaneous distribution of causal influence. This is important since change of position is a paradigm case of succession; so succession is grounded by simultaneity of causes. Third, for the chain of *previous* experiences to 'cohere', it has to have one single result in the present, conserving the whole causal world together at each moment. The paradigm where a ball makes a causal impression on the cushion that exists simultaneously with it and the paradigm where two successive movements of billiard balls meet simultaneously when they hit, both require coherence across changes of place and state. The movement of the boat through the river presupposes world-simultaneity just as certainly as do the rooms in the house.

Kant's operations of coherence, coexistence, chains of reciprocal causality, and communities of interpenetrating substances, sound Leibnizian. This is not necessarily bad, but it throws simultaneity back into troubled waters. In Leibniz, every substance reflects, from its own perspective, all the determinations of all substances, past, present and future. By Kant's definition, this would make all substances simultaneous. Leibniz's proviso that each contains the others from its own perspective is supposed to save each substance's particular time in the order of succession. But it is not easy to save temporal individuation on this model. For if my substance reflects your properties according to my perspective, then my substance reflects your perspective on my perspective on your perspective, to the point where it is not clear how to distinguish my perspective from yours. Similarly, if I reflect the simultaneity of world-predicates from my position in temporal succession, but my perspective of succession is reflected in simultaneous perspectives, it is not clear how to individuate moments in time.

Kant did not intend to absorb succession into perspectival simultaneity in this way. It seems that a community ought to be able to refer back to its simultaneous parts without the whole collapsing into a single moment. To work out the economy of communal return, and to see how individuated succession might be restored within general simultaneity, we turn to Hegel.

(b) Simultaneity in Hegel: The System of Expanding Explications: Succession via Divisible Simultaneity

Hegel retains from Kant the idea that simultaneity depends on going back. But instead of the subject looking back and constituting simultaneity, Hegel makes space and time themselves go back and create simultaneity.

Hegel's discussion of simultaneity begins his *Philosophy of Nature*[7], the middle book of his *Encyclopedia.* The first book, *Logic*, ends when it has described all of the formal categories that something has to instantiate in order to be. Once something instantiates all necessary categories (once it has qualities and quantities, causal relations, etc.), then we can say that it just is. The thing is spread out, *Dasein*, being-there. Insofar as reality has the status of being blurted out, we call it space, Hegel says. Each part of space is simply laid out alongside the other parts. Nature, the region of being that is just there, starts with space. And the first characteristic of nature is simultaneity: the simple being there together of all that is. As we will see, the differences that emerge in being laid out, that is, in space, that is, in simultaneity, introduce change, organization and reorganization, movement, and successive time.

For Husserl and Kant, simultaneity is built up out of succession; for Hegel, succession will arise from the collapse of simultaneity. The activity of simultaneity is succession. There are two key points: (a) 'The truth of space is time, so that space becomes time' (s. 257) and (b) 'The point behind [on a circle] is at the same time that at which it will arrive' (s. 261). The first defines space dynamically, and the second defines simultaneity as the temporal dynamic of space.

Nature is self-externality. It is 'collateral' (s. 254): lateral, with things beside each other, and *co*-lateral, because everything reaches everything en route. Space is not originally made up of points, but continuous. We tend to imagine that points are external to one another, but externality is dynamic continuity before it is discrete isolation. For Hegel, a point arises only by negating the rest of space, and resisting the incursion of the already continuous spatial field (s. 254). Divisions are constructed out of negations within a continuum. We tend to think that the point is the first division of space, but if space is primordially continuous, it is just as plausible that the first division in space is a line, or a triangle for that matter, or a plane or volume. Similarly, there is no ultimate meaning to up and down, or side to side, in a world that can be rotated. The whole of space is first, and divisions secondary. Saying that space is divided is not wrong, but it divides freely.

Now, although division is secondary, it is also real, so divisions, for example, up and down, extend throughout the continuum. Any given point into which space divides is both *this* point as well as the lines that extend outwards from it in all directions. If one starts a theory of space with points, then it would be a metaphor to say that points 'extend' into lines, but if one starts with continuity, then a point is the result of contraction and is already literally continued into lines. That a line is extended from and through points does not mean that each here continues elsewhere, but the line is continuous in the sense that the series of extended points is unbroken. When a point negates space, it simultaneously extends through space (s. 256).

A discontinuous point both negates continuity, and by building a line, simultaneously negates every other (equally negative) point — it negates the negations. And because the point negates continuity by negating a discontinuity (i.e. by negating the other point), the result is discontinuous continuity — a line made up of points that cancel each other one after the other. A line is a specific portion of space built out of cancelled points. The main consequence is that any given point can lead by some direction or other to any possible line anywhere in space. That all points lead towards all lines implies a plane, and planes rotate through points into volume. So even though each division is arbitrary, the sum of division is not arbitrary, because it leads itself back, multiplying itself into the next point until it fills up space again. On the one hand, the point extends out towards another in a particular direction; on the other hand, it is possible that it could go in the other direction. Indeed, it can go in one direction only *because* it is also possible that it could go in the other. In a virtual sense, it *does* also go into those other directions at the same time.

Hegel does not agree with Kant that the proposition, 'The shortest distance between two points is a straight line' is synthetic. Hegel says it is 'analytic' that shape unfolds into externality, following a rule of 'simplicity of direction' (s. 256). Analyticity is more dialectical than externally conjoining two things in synthesis.

These properties of space generate theorems of simultaneity. Simultaneity is, in one sense, simply the fact that all space exists at the same time. In a second sense, simultaneity consists in the way that lines are possible because other lines could have been drawn at the same time, and virtually are also being drawn at the same time, extending in turn into additional points that will also be there at the same time. The first sense is the spatial definition of simultaneity, the second is its temporal definition.

'The truth of space is time, so that space becomes time' (s. 257). In brief, the argument is that (a) any finite location is a space that has negated the 'being-for-self' of space, and so sets itself apart and encloses itself; (b) self-enclosed spaces are indifferent to each other; (c) the distance between spaces is not determined inside each space's borders but by the movement from one to the other; yet (d) it makes a difference where one location is, relative to others; therefore (e) the difference between locations is a time difference as much as a space difference. In simple terms, space is differentiated, but it only makes a difference which point of space is where, if it is possible to pass through a point at a certain point in time. It is not that Hegel brings in the *ex machina* assumption of a subject deciding where to go next in space. His argument is that self-enclosed parts of space exist for themselves, and just for that reason, space is 'no longer paralyzed' (s. 257). A self-enclosed space is the sort of location that *can* be occupied, that *can* make a difference to where something is, and just because of that, space is the medium for mobility, for going from one space to another, for changing place over time, for location to take place in time, for 'the truth of space to be time'.

On the one hand, the quasi-temporal character of spatial mobility re-establishes the overall simultaneity that self-enclosed spaces threaten to fragment. Spatial extension, beginning from a given spatial point, is an ontological movement that reproduces that point in a successor, traces points into a line, shifts lines into planes, and inflates planes into volumes. Each separate spatial being negates the context of multiplicity in order to maintain its determinacy. That is, space is the sublation of abstract points in directionalities and coexistences. In simple terms, a spatial point is what it is only because it is placed beside another. But while space in this way has the structure of successive points along a direction, it depends on the multiplicity of places being given together and *not* awaiting a future movement. Indeed, if each point in space had to be seen at a separate time, or indeed to exist at a separate time, that is, if each point had its neighbours not at the same time but in a next time, then there would be no space at all, only time. Space is 'immobile collaterality' (s. 257). But the same logic that entails that each spatial point must have neighbours in every possible direction also entails that a point in space cannot be more tied to a neighbour in one direction than one in another direction. This indifference of each point relative to each other is essential to the simultaneity of points. For it is only if each direction on a plane is always a possible continuation of any point that all points on the plane exist at once[8]. The drawing of any line can always be broken off just because other lines are there at the same time.

On the other hand, this indifference of place to place implies freedom of movement. And freedom of movement frees temporal succession from spatial determinacy. It is because space breaks up indifferently that points in space can be traversed over time. It is in this sense that the 'negativity' of space is time (s. 257). Even though simultaneity is spatial coexistence, it is the time-form of space: space as the not-yet-altered coordinate, the alterable, or the already-altered-and-already-recoordinated. Manifold simultaneity is the kind of space within which something can happen in time.

A point does not actually move over into the next point. But points are relay junctions on a field in which it makes a difference where each point is. This difference that makes a difference is the temporal character of space.

Time is the negative character of space, that character of the point that negates the spatial continuum in favour of a particular location that makes a difference. Time is not really a dimension on its own; it is more the negation, or the freedom, or the mobility of, spatial dimensions. It is not for nothing that some philosophers deny that time exists; yet without existing, it has a logic of its own. Alluding to the *Phenomenology of Spirit*, which argues that a moment in time is not even simultaneous with itself, Hegel says that 'In that [time] is, it is not, and in that it is not, is' (s. 258). Time is the form of negativity relative to any space (s. 257). Time is abstract, not a positive part of nature (s. 258). But the fact that time is the negativity of any and every unique point or enclosure insofar as another point can adjoin it implies that time is the *point-by-point* negation of points of space. Time therefore also has its punctual or 'momentary' concretions (s. 258). Time can be divided, just because time is that which divisible space becomes.

Like space (s. 258), time is continuous precisely when its moments arise and pass away. Like space, time can be abstracted from its moments, as if it were a container. That is, like space, the form of time is a priori, the unsensed form of sensibility. But also like space, time is not a mere container of the sensory field; it is the real material of becoming. Time *seems* like an abstract container, and seems to be eternal, only because it is universally the case that things change over time. Like space, time therefore incorporates the opposition between universal form and unstable differentiation. Time is the negation of momentary difference in universal becoming. But unlike space, which is *subject* to the negation of the field by its parts and vice versa, time *is* that negation.

Precisely because time is the negation of moments, it exists nowhere but in individual differentiated moments here and now (s. 259). Each here and now is 'finite', 'affirmative' and 'fixed' (s. 259). Past and future do not exist in nature, since natural moments are here and now and nowhere else. From this classical position, Hegel draws an odd original conclusion: 'the past and future of time are space' (s. 259).

It is plausible enough to say that the only time that ever exists is the now (and its vanishing). Past and future are not other nows waiting their chance. If a second time also exists (past or future), it will have to exist in the now as well. Past and future may have a temporal status different from that of the now (as Deleuze says), but they can exist nowhere other than the now, albeit differently from the way that the now exists. Other moments in time must exist at the same time as the now. It is in this sense that Hegel says, 'the true present is therefore eternity' (s. 259). This is not a *totum simul* in which nothing changes; on the contrary, change has its status in something other than the here and now that exist in each present. In the form of memory and hope, for example, past and future consist in the fact that something other than natural change takes place in the now, namely, reflection. In intersubjective form, past and future consist in the fact that people other than ourselves, elsewhere in the world, have projects at the same time we do. And in physical form, past and future are implied by mobility across the spatial plenum. Where I have been in the past is somewhere else, and that somewhere else exists now—my past took place somewhere else, where something else is happening now. The past and future of *this* here and now are referenced to possibilities in the now of *other* heres. The status of other times cannot be exhausted by what is retained and anticipated at any one moment in space, but consists also in what other things are mourning and hoping for, reflecting and preparing for, elsewhere in space at the same time.

The principle that the past and future of time are space makes use of the isomorphism shared by the spatial here-point and the temporal now-moment. Everything past, present, and future must exist in this now, and this now exists in points of space. A temporal moment must exist as the present, as the has-been, or the still to be, of a here. Furthermore, all time is in points of space, not limited to this point but distributed throughout the past- and future-related reflections present in all points of space. In this sense, Hegel says, just as space revealed itself to be time (since each point in space has other spatial locations as the moments that could make a difference for it over time), so time reveals itself to be space (since a next time can only actually exist as what exists in space) (s. 260). Future time has to come into existence across the full continuity of all space at once. But the 'all' of space that constitutes the future has to be distributed, so that each point in space contains its own share of the future of each prior here and now. Space depends on the future being shared across points, and time depends on travel across space during a succession of moments. Each spatial point must be determin*able* : it must have a time-river run through it; otherwise, it will not be differentiated from the spatial continuum.

The simple version of all this is that if there are different spaces, it must be possible to move through them, so a multiplicity of space implies time; conversely, movement during the passage of time implies that space persists as long as movement does. The parts of space are not only laid out simultaneously; in addition, spatial extension is where temporal succession occurs. This kind of 'concrete point', equally spatial and temporal, has a traditional philosophical name: 'place' (s. 260).

Hegel is deliberate about defining space and time together. 'This here is at the same time a now'. And in the movement from one here to another, 'This here is to the same extent time... or a now which has been' (s. 260). A place is temporal enough to exist simultaneously with the time in which something passes through it. If space and time could not be defined together in place, it would make no sense to say that a point of space is simultaneous with anything. But Hegel's thesis is that time exists as soon as, and only once, spaces are different from one another. This means that there is no such thing as temporal succession in which time passes in one solitary point of space but not in another, and there is no such thing as a universal space that exists without change of place over time. Of course, one can stay in the same location in space for as long as one wants, but it will not be 'the same *place*' once the situation has changed and it has ceased to be the hotspot. Similarly, one can endure over time for quite a while, but doing so will not get one closer to the end of a waiting period unless one has held one's place in line. In place, that is, in context, the here exists at the same time as the now. And if places are arranged in succession relative to movement (whether it is the actual movement of matter or the movement of subjective attention), then space is a system of possible temporal successions. Space is the simultaneously laid out routes for successions.

When something goes to or from a place, it takes up a space at a time, its proper place or a wrong place, a space that matters. The right space to be in, comes into being and passes out of being in time; being there at the right time is a there in space. Space and time thus 'regenerate' in each other's points (s. 261). And when an arrow moves, or some other process takes place, the place no longer makes the same difference; the time for it has passed, and the appropriate place to be has moved on to some other point. When place is thus defined as changeable space over time, place is 'motion', the motion that gets to or from the target space at the time. When time measures movement from place to place, movement is 'matter' settling into place (s. 261). In Hegel's philosophy of nature, matter is not raw material but the situated mobility of what is in place at the time.

We can now see the shape of simultaneity on the model of place. Since matter is not indifferent to the place it is in at the time, it constitutes an arrangement. Matter is a composite 'being-for-self' (s. 261) in a space-time environment, a restricted simultaneity. Yet since matter is at one place at a time, it is 'impenetrable'; each piece excludes other pieces of matter, each is 'singular' (s. 261), a negative simultaneity. After all, simultaneity refers to some *other* thing existing at the same time. Two things that exclude one another can be simultaneous, yet two things that (if this were possible) were so distant as not even to exclude one another, and could not possibly exchange places, would not be simultaneous.

The being-for-self of matter in movement in a singular place is the culmination of Hegel's account of simultaneity. Simultaneity is congealed movement in matter. Motion is 'implanted' in matter by the physics of 'force', as when equal forces on two sides of a lever hold two material objects up simultaneously, or when a brick is thrown towards a man's head with a force sufficient for the brick and the head to share a moment in the same place. To follow Hegel's construction of physics one step further, simultaneity consists neither purely in the layout of space nor purely in the instant of time, but in the relation between soliciting and solicited forces, that is, in the masses that attract one another into being in the same place at the same time, namely in gravity (s. 261). Gravity plays a crucial ontological as well as metaphorical role in the 'idealist' philosophers of the period due to the way cause and effect are reciprocated therein. Like the paradigms of the two poles of a magnet, or the sender and receiver of a signal, gravity is a paradigm of concrete simultaneity—not abstract contingent simultaneity but concrete simultaneity based on what Kant calls the interpenetrating determinations of substances.

Hegel thus fleshes out the objective standard of simultaneity that Kant was aiming at. But by not separating the simultaneity of objects from subjective movement through space, as Kant did, Hegel preserves resources for objectively individuating temporal perspectives. For Kant, if one can return to the place where an object was once seen, that proves that the object was there simultaneously with other objects seen in the interim. The problem was that other things might have been happening to the object during the interim. Kant does not mind if the test for simultaneity occasionally gives a false positive, since the structure of simultaneity is a priori while the actual simultaneity of two empirical objects is only known a posteriori. But if there is no mechanism for repeating, refining, or verifying the test results, then Kant's argument for the simultaneity of substances is undermined. Hegel's solution, common sense as always, ties the test of simultaneity to the ontology of forces rather than to intuitions. For Kant, 'return' means that a perception is repeated; for Hegel, 'return' means that a cause is still operating through its effects. Simultaneity is tied not to space in general still being there but to particular matter in motion. Hegel's twist is to find the ground of simultaneity in the cause-effect relation, the very relation where Kant found succession instead. The impact of this difference is that for Kant, simultaneity consists of things in two *spaces* existing at the same time. The paradigm is not so much two things existing at once but one big thing (like a house) spread out. For Hegel, simultaneity consists of things in two *places* existing at the same time. The fact that an observer walks back and forth from room to room differentiates the environmental space between them, that is, in terms of their place. If the rooms are simultaneous, it is because they are capable of accepting the arrival or departure of the body passing through the door. The difference between a person who moves from room to room and a sailor who might turn the boat around, is not crucial for defining simultaneity. Simultaneity is found in the departure from one place and the return to the same environment in another place. The simultaneity of the environment is verified by creating different scenarios, rather than by rediscovering the same scene. The paradigm is two differently functioning things in some phase of interaction.

Of course, Kant does not deny these layers of simultaneity—indeed, he insists on them in causal community. Nevertheless, Hegel's paradigm of simultaneity focuses not on self-same motionless things that a spectator can look back and forth at, but on motions that make things different. It cannot quite be that times simultaneous with this one take time to get to. But what makes events simultaneous is that their times overlap in the form of difference, not in the form of sameness. Once we question whether two events operating in different rhythms can exist exactly in the same points of time, the only way to avoid an outright denial of simultaneity is to conclude that simultaneity consists of the way that different time-flows overlap.

Of course, we expect from Hegel that every activity should in some sense be simultaneous with every other. Each movement from room to room, or up and downstream, constructs the simultaneity of houses and landscapes. But a new problem emerges if simultaneity is not given with space but waits to be constructed over time: What is behind the construction of those simultaneities that are *not* actually constructed by the force of some movement? Is the simultaneity of those rooms that are not yet entered delayed until such time as someone or something crosses the threshold? What is the status of the simultaneity of pathways not taken?

Hegel's solution involves rotary motion and the spokes leading back from a circle to its (missing) centre.

Let us begin again with continuous space: 'One place does not merely imply another, it sublates itself into becoming other' (s. 261). Zeno's premise is that a flying arrow is at rest at each moment at each point of space. Hegel's thesis is that each place *becomes* the next place. The arrow may be in one space at a time, but it is not at one place at a time, precisely because the place (the space that matters, i.e. the state of a changing situation in a space at a time) already occupies the next space as well as its own space at the same time. All the arrow has to do in order to move is to ride the wave of place. Of course, there has to be a force that moves it along one such wave rather than another. But for Hegel, much of movement is topology, which shapes the succession in the arrow's positions.

For Hegel, the paradigm of position-shifting is circular motion. Rectilinear motion ultimately has a similar structure, but the latter's connotation suggests a one-at-a-time transposition to another point. Circular motion suggests returning to points already occupied. Something remains in the places the circulating thing was occupying a moment ago. 'The point behind is at the same time that at which it will arrive' (s. 261).

Ultimately, simultaneity consists in the fact that movement is the 'return' of a line, the 'reestablishment of the immobility of space' (s. 261). On a circle, the future is behind us, therefore 'the goal is past, not future' (s. 261). Yet as we have seen, place is affected by the things that move into it, and space is eternal only by being topologically altered in time. So in Hegel's metaphor, the circle dissolves its own centre. We return precisely by *not* being able to get back to exactly the same place. The circle (and any line) preserves its last point and returns to it, by cancelling the latter's presence, that is, by getting to the next place at the same time. The better image of Hegelian space-time is less a circle with an empty centre than a ring expanding outwards in concentric circles. In an expanding universe (and for Hegel, even a universe that does not increase in size is expanding in the sense that points extend into volumes), all the points maintain their simultaneous relations even while they are repelled over time. The circumference of the expanding circle touches more and more space and time while gradually, but never completely, escaping the gravitational pull.

But there is one more feature in Hegel's image of simultaneity: in the expanse of circular motion, it is always possible to make an abrupt right angle turn back from the periphery into the heart of the circle, to cut through the wave, either to open up a pathway to a different set of simultaneous objects, or to break the simultaneity revealed by repeated natural or habitual cycles and introduce new distinctive successions into world-history. It is not the centre but the attractions and repulsions around it that make space. A straight line is the spin-off that leaves the orbit of a rotating circle (ex-orbitant, as Derrida would say). This answers the question above as to how Hegel constructs the simultaneity of those objects that we do *not* normally return to. Recycling a single set of points would not do it. We have to cut across one set of simultaneous points and enter other sets, cutting lines through circles, changing course in mid-stream, breaking through walls to see what is between the rooms of a house, multitasking epistemically. Breaking a cycle of simultaneity means engaging in one-off lines of enquiry, that is to say, engaging in succession. But each new line in turn establishes a simultaneous set of positions of its own. The succession of lines of simultaneity constructs a power set of sets of simultaneous points. By cutting across different simultaneities, we preserve the simultaneity of sets of objects whose mutual interaction has been delayed. In Husserl, we needed multiple rhythms of succession to constitute simultaneity. In Hegel, we need multiple sets of simultaneous objects in order for simultaneity to be constructed in actuality yet remain virtually unbounded. For Hegel, this is how simultaneity generates succession.

Perhaps this is not so different from Husserl. A subject coordinates processes into rhythmic equilibrium. I start chewing in sync with my walking, pulling all my activities into the same circle. To coordinate the time-flow of different anticipation and projection schemes, I cut across rhythmic cycles.

For Hegel, succession occurs when one cyclical system cuts across another. Linear time cuts across repetition, and so it is more multi-layered than circular time. In linear time, an event does not just carry on along the lines of its previous moment, fulfilling anticipations and retentions. Linear time cuts across different rhythms of retention and protention. Time is linear because it contains simultaneous successions, because simultaneity is cut across by ways of returning other than those that have yet been carried out, because a circle or a habit can be cut otherwise, and because such delays are built into topology.

Kant's idea that we can subjectively go back to what we saw another time is not enough to construct simultaneity; his idea that things occur in causal interaction constructs objective simultaneity, but at the Leibnizian cost of underplaying succession. Hegel's dynamic space-time builds simultaneity and succession into space and time together. Dynamic movement makes simultaneity by forcing subjects to go back to something that was *not* there before, something that was virtually built into objective space and time but never before actualized. The dynamic posits a subjective pathway through objective space-time: virtual world-simultaneities that coexist due to the linearities that traverse concentric motion.

But we need to say more about simultaneity not travelled: virtual or delayed simultaneity. Delayed simultaneity must be virtual, yet real. Otherwise, there is only a longing for simultaneity, in Fichte's sense.

Fichte's three-stage dialectic (*Science of Knowledge*[9], 203), in which the ego posits itself, then posits the not-self, then posits itself as limited by the not-self, can be read as a three-sided dialectic of simultaneity. (a) The ego's unrestricted activity posits general coexistence. (b) When the ego's self-projection forms and hits target objects, its cognition reflects off each object towards others, rendering them all individually simultaneous. (c) But each object resists the ego's activity, and thereby limits which objects appear simultaneous with which.

Schelling, during his Fichtean period, declares that, 'All simultaneity occurs only through an act of the intelligence, and coexistence is merely the condition of the primordial succession of our presentations'[10]. Where Schelling's later absolute idealism, like Hegel's, expresses objective simultaneity as a kind of ontological gravity, Fichte's subjective idealism expresses simultaneity as property. Each subject marks a range of objective reality by seizing it as private property, recognizing other subjects' capacity to seize other parts of the world in the same way, and then signing a social contract that establishes simultaneous subjects[11]. Fichte's *Science of Rights* is notorious (thanks to Hegel) for supposing that the recognition of other subjects requires police surveillance and passport identification checks. Perfect simultaneity cannot be actually achieved. But Fichte says we yearn for it, and so we are tempted to demand more results than we can acquire. For Hegel, in contrast, the construction of simultaneity can be neither rushed nor deferred indefinitely. We need an ontology of virtual simultaneity, so that its delayed construction can still count as an achievement. For this theory, we will turn in the next chapter to Bergson. In addition, we need a phenomenology of the construction of delay, so that experience can force simultaneity out of delay, and not just wish for it or claim it against the evidence. For the plastic achievement of simultaneity in the face of delay, we turn to Lessing.

(c) Delay in Lessing and Historical Dialectic: Lingering and Repetition

Kant's back-step simultaneity and Hegel's cut-up simultaneity provide virtual epistemology and ontology for the simultaneous world. But the phenomenological and experiential representations of that simultaneity remain problematic. The background simultaneity not yet reached by backward movement and the virtual simultaneity that is invisible to embodied locations remain mere expectations, albeit rationally grounded.

The imaginative capacity to see movement before its time, and to see the end of time while it is still moving, plays out in Enlightenment progressivism from Kant's 'Perpetual Peace', through Hegel's philosophy of history, even to Marx's spectres of communism to come.[12] Any idea that the future is destiny has to explain why it has not yet arrived. The experience in which simultaneity is delayed by succession, succession stalled by simultaneity, and delayed succession achieved as reality in art is prevalent in pre-Kantian philosophers like Lessing.

Aesthetics since ancient times has tried to arrest movement[13] in non-moving representations and to represent motionless images in moving media. For Lessing[14], the divide between movement and non-movement differentiates plastic arts like sculpture, which employ signs simultaneously, and literary arts like spoken poetry, which employ signs successively. Art forms ought to depict the kinds of objects that suit the temporality of their signs: sculpture should portray beautiful bodies at their best moments, and poetry should portray dramatic problems and their resolution. Lessing is criticized for exaggerating the dichotomy, but in fact, he is most interested in hybrids, like the ancient statue of Laocoon.

Laocoon, the priest of Apollo in Troy, interrupted his prayers to Apollo so he could warn his countrymen not to accept the gift of the horse; Apollo naturally then sent serpents to drag Laocoon and his two sons into the sea. The sculptor would like to depict Laocoon in full scream, simultaneous with the serpents, the sons, and the drowning, but he cannot do so, since that would distort the body's beauty. Such pain does not deserve to be visible. The sculptor can, however, arrange that Laocoon's scream be experienced: he can present the viewer with the scene just prior to the scream and make use of the viewer's natural tendency to imagine the scene that follows the one he is presented with (ss. 1–3). The sculpture provokes the viewer into articulating the successive events of the narrative. He delays the scene by showing its predecessor instead as a phase of succession. It is not that a succession has its result delayed but that a simultaneity is delayed by succession, which intensifies the simultaneity imagined but unseen. For Lessing, this greatest of sculptures lingers at the transition between simultaneity and movement, between stasis and kinesis.

Goethe's essay 'Upon the Laocoon'[15] offers a different solution. Instead of the viewer lingering over the sculpture as a motionless object, and adding movement in imagination, Goethe suggests looking at the statue as if it were moving—as if a moment ago the figure had been serpent-free, and a moment hence will fall into the sea. Only if it can fall dead, can it 'present a fresh, living image'. Goethe's technique aims to trick the eye:

> In order to conceive rightly the intention of Laocoon, let a man place himself before it at a proper distance with his eyes shut; and then let him open his eyes, and shut them again instantly. By this means he will see the whole marble in motion; he will fear lest he finds the whole group changed when he opens his eyes again. It might be said that, as it stands, it is a flash of lightening fixed, a wave petrified at the moment it rushes towards the shore. The same effect is produced by the contemplation of the group by torchlight.

The static image is framed by a flash of light. The eye is deceived into thinking that the statue is a slice of ongoing motion. Tricked into projecting a rapid succession of before- and after-images around the stop-action statue, the viewer is treated to cinema and does not need to add his own screenplay.

Lessing rejects the idea that a series of frames amounts to narrative. He ridicules the suggestion of a certain Count Caylus that a good poem would fill a gallery with paintings (s. 13). A viewer could never run around a gallery fast enough for the frames to blend into a cinematic viewing of the original poem. A visual sign can, however, provoke the viewer into narrating a history. Indeed, many eighteenth-century aesthetic phenomena, from marionettes to tableaux vivants, blur the border between static and kinetic.

The dispute comes down to the silent scream. It is not actually clear to me whether Laocoon in the sculpture is screaming—Lessing thinks he is obviously not; Goethe thinks he plainly is. But the principle, for Lessing, is that if a viewer were forced to look at Laocoon actually screaming, his compassion would soon be deadened; the viewer has to do the screaming himself, to do the statue's moving for it. Although death is the ultimate still life, moving art like poetry can bring death back to life.[16] The viewer's imagination acts like poetry, temporalizing the plastic body. But what in the physical sculpture moves imagination? The sculptor's technique is to carve the subject in an incongruous position, one limb in a position it must already have vacated, another in a position it could not yet have reached. The limbs on a sculpture exist simultaneously, but the contradiction in simultaneity implies succession.

By the converse principle, the words in a poem are read successively, so when depicting a simultaneously existing complex, like a face, the poet contradicts his medium. Written straightforwardly, any poem about a face is ugly. No matter how beautiful the words, it describes the face one organ at a time, which is grotesque (s. 17). One solution is for a poet to narrate the successive actions of a painter painting a face (s. 19). Homer's verses on Achilles' shield do this. Virgil does not do as well; he merely describes the pictures juxtaposed on the shield. Homer describes the painter painting those pictures over time. But this is still not enough, since the poet can show the painter too only through motion, never in the restful contemplation of beauty. A second solution also comes from Homer: Helen's beauty is sung by describing her affects on the Trojan elders (s. 21). Poetry reflects beauty through its lustful and nostalgic reception at the stiffened hands of old men. But waiting too long for the viewer is as bad as not waiting at all. If poetry is an old man's reflection on beauty, when does the youthful beauty of vigorous movement actually take place? Ultimately, for Lessing, movement becomes visible, and succession makes simultaneity visible, not through the gimmicky speed technology in magic lanterns, and not through modern art's obsession with 'hyperbolic' movement 'divided and scattered', as Winckelmann put it in 1755,[17] but by lingering over figures like Laocoon about to scream. Goethe tries to define art in terms of continuous succession, but the movement is short lived; he sees the movement too soon so it freezes up. For Lessing, it is 'lingering' that gives movement its history[18], and makes coexistent and successive signs coexist in the successive ages of seeing and writing. The Greek thereby becomes modern, painting becomes poetry, and delay finally becomes simultaneity, albeit virtually.

For Hegel, simultaneity is a condition for unfolding succession, just as for Husserl, succession is a condition for contraction into simultaneity. In general, idealism introduces lines of freedom into simultaneity—the freedom of geometrical possibility, the freedom of indifferent development, the freedom to move through a gravitational field, the freedom to resist incursion, and the freedom to mark one's own region of space at the same time as others mark theirs. The danger is that if simultaneity constitutes the freedom in which all things are possible at once, it may revert to a pre-given space in which possibilities are all already laid out. We cannot solve this problem simply by saying that succession allows events to expand freely. Unless succession preserves the contemporaneity of events, it isolates and disperses events altogether. Simultaneity is too tight, but succession is too loose. What is needed is the level of simultaneity that I am calling delay.

Hegel's philosophy is filled with multi-layered, cross-appointed, interfering, field constructing activities. I turn next to Bergson's theory of the virtual connections that constitute multiple layers of simultaneity existing simultaneously in a dis-synchronous overlap of flows.

As Bergson's *Duration and Simultaneity* takes Einstein as its point of departure, we need first to touch on simultaneity and delay in the sciences.

Chapter 5

Simultaneity and Delay in the Sciences

Theme of the chapter: Simultaneity is calculated relative to frames of reference. The problem of synchronization could lead to conventionalism, to denying time passage altogether, or to a Bergsonian disjunction in the time-line. Cognitive science finds many staggered micro-series without a central point of judgement and posits a delay-function in consciousness.

Philosophers need to reflect on phenomena drawn from the sciences. Non-experts like myself who speculate on these fields will make mistakes and oversimplifications. But what choice do we have?

(a) Simultaneity Relative to Frames of Reference: Physics

I have been developing the theme that simultaneity is built out of divergent time-lines. Simultaneity is not a form of identity, where two things happen in the same instant, but is rather found in the juxtaposition of different time rhythms.

For Einstein, the judgement of whether two events are simultaneous depends on the frame of reference from which they measured. There is no one overall frame of reference, therefore no universal way to decide whether two events are simultaneous. The theory of simultaneity has to take into account the divergence of time-lines, or world lines.

Einstein would not say that time is the moving image of eternity, that time is an illusion, or that the truth of events is outside the passage of time, all taking place at once in the mind of God. But some interpretations of relativistic time, influenced by McTaggart, take events to be mapped together on a diagram of space-time all at once. John Leslie, for example, thinks Einstein shows something like Spinozistic eternity.

At a general level, classical science assumed that a causally interactive world, though changing, is at each moment all there at once. But on specific points, as Max Jammer's *Concepts of Simultaneity: From Antiquity to Einstein and Beyond*[1] shows, the history of physics is full of challenges to the universal simultaneity of the now. In ancient times, for example, Sextus Empiricus rejected astrology on the grounds that we cannot verify the position of distant stars simultaneous with the birth of a child (21). In modern times, Ole Roemer discovered (in 1676) that light has finite velocity, that is, that 'distant simultaneity' is measured by taking into account the 'retardment of light'. It follows that the 'universal simultaneity thesis', namely that what we see simultaneously is objectively simultaneous, is false: 'the world as we see it is an optical illusion' (295–6). Still, this kind of delay can be calculated and does not imply relativity.

Newton knew that motion is relative. In a universe of only two entities moving past each other, there is no way to determine which is moving and which is at rest, or whether both are moving. But Newton did not anticipate Einstein's argument that time is relative as well.

The starting points of the special theory of relativity are that the speed of light is constant and that the laws of physics, including the speed of light, operate in the same way everywhere in the universe. Peter Kosso[2] points out that 'relativity' is thus a misnomer. That the speed of light is constant is strange but true. The speed of other things is not constant but depends on the medium that the thing moves in. If we toss a ball so it is moving five miles an hour on the ground, assuming the ground is not moving, the ball moves five miles an hour. If we are in a train moving fifty miles an hour relative to the ground and we toss a ball five miles an hour relative to the train in the direction that the train is moving, the ball moves fifty-five miles an hour relative to the ground. If we throw the ball backwards inside the train, it moves forty-five miles an hour relative to the ground. The speed of the ball is not absolute but is measured relative to a frame of reference in movement or at rest.

Strangely, the speed of light does not work like that. If we shine a light ahead of us while we are motionless on earth, the speed of light is 3×10^8 meters per second. If we shine a light ahead of us while we are moving forward at a speed of thousand meters per second, we would think that the speed of light would be 3×10^8 *minus* our speed (since we should be catching up to the light), but it is not. Similarly, if we are moving backwards and shine the light forwards, the light still moves ahead of us at the speed of 3×10^8. No matter how fast we go in any direction, light moves away from us at 3×10^8 meters a second. It is constant: unaffected by the speed of the frame of reference in which it moves.

Special relativity defines a frame of reference as an inertial system of bodies; all the things within the frame (like a train) move together (simultaneously, in what relativity considers an unproblematic sense) at a constant rate of speed.

Einstein's best illustration for the layman of the relativity of simultaneity involves a train moving past a station: the train and the station are two frames of reference moving at different relative speeds.[3] The measurement of whether two events are simultaneous depends on the observer's frame of reference, that is, on the observer's motion relative to the events. Suppose that as a train moves by the station, lights go off at the front and back ends of the train. The light coming from the front end and the light coming from the back end meet in the middle of the train. A person standing *inside* the train, at the mid-point of the train, sees the two light flashes meet her at the same time. That she *perceives* the two light flashes simultaneously is considered by relativity physics to be unproblematic. The problem arises when she calculates when the two lights must have been turned on: given the speed of light and given the length of the train, did the two light sources flash at the same time? She knows they reach her at the same time, but she has to calculate whether the two lights at the two ends of the train had flashed on at the same time. Meanwhile, another person is standing *outside* the train, on the station platform, which is not moving. At the moment when the two flashes reach the person on the train, he is standing exactly across from her. The guy on the platform sees the two flashes reach him simultaneously, just as the person on the train does. (Let us not say that the two people 'simultaneously' see the flashes; the thought-experiment need not assume that, and it would raise problems.) Like her, he *perceives* the two flashes simultaneously, and like her, he has to calculate whether the lights at the two ends of the train had flashed on at the same time.

Consider the calculations from the two points of reference. The person inside the train knows she is at the mid-point of the train and knows that the speed of light is constant. The two flashes of light left the ends of the train, travelled the same distance at the same speed, and meet simultaneously at the mid-point of the train. She must draw the conclusion that the flashes of light left the two ends of the train simultaneously.

The person standing on the platform watches the train pass. He sees the two flashes arrive at his position simultaneously. But he sees the train moving by, let us say to the right. Since the light beams meet when the mid-point of the train is exactly in front of him, the light beams must therefore have left the light sources at the two ends of the train while the train was still to his left. The two ends of the train are equidistant from him when the beams meet, but when the lights first flashed on, the train had not reached him yet, so at that moment, the front of the train had to have been closer to him than the back of the train. So since both lights meet *him* at the same time, the light flashed from the front end of the train will have travelled a shorter distance to get to him than did the light flashed from the back of the train. We know the speed of light is constant. So if two light flashes are travelling at the same speed and one travels farther, it requires a longer time. The two lights meet him at the same time, so the light from the back of the train must have been travelling for a longer time. Therefore, light flashed from the back of the train must have left its light source *earlier* than the light from the front of the train. So the two lights did not flash on simultaneously! The person on the platform has to say the two lights flashed on at different times. He makes this calculation accurately, based on the position he is judging it from and the movement of the train relative to him. The person on the train, just as accurately based on her frame of reference, calculates that the two lights flashed simultaneously. One calculation says the two events are simultaneous; the other says they are not simultaneous. Both are correct calculations based on their frames of reference. There is no universally correct judgement regarding simultaneity; there are two correct calculations depending on one's frame of reference. If there were an absolute way to determine which of the two frames of reference was the correct one for measuring speed, we could say that one of their judgements regarding simultaneity takes precedence. But since motion is relative, we cannot say that either the train or the platform is the ultimate frame of reference for determining movement. Their differing judgements regarding the simultaneity of the original light flashes are equally valid.

The relativity of simultaneity is not relativity of perception but of measurement. From ancient times, philosophers have known that judgements about things are relative to perception. Everyone knows that when we experience lightning, we experience the flash of light before the thunderclap. Our experiences of the two aspects of lightning are successive and not simultaneous. But once we know how it works, we calculate rightly that the light and the sound originated simultaneously. In classical physics, we can perceive two things non-simultaneously but calculate the objective truth that they occurred simultaneously. In classical physics, the perception is relative but the objective fact of simultaneity is not. Einstein proves that it is the objective calculation of simultaneity, and not just our perception, that is relative. Two events actually *are* simultaneous according to one correct calculation and *not* simultaneous according to another.

Objects moving close to the speed of light undergo measurable time dilation. Try measuring the subatomic particles called muons that come through the atmosphere at close to the speed of light: depending on whether you are travelling with the muon or standing on earth, the muon exists for a different amount of time. I have not performed this experiment myself.

If we put a clock on an object moving very fast, then according to the calculations of a person at rest, it will seem to be ticking more slowly than an equivalent clock at rest. A person who is moving at the same speed as the clock thinks his clock is ticking normally. Imagine a clock that sends a flash of light to a mirror, so that the light is reflected back to it; at the moment the light returns the clock, it ticks off one second. When we are at rest relative to our clock and its mirror (or moving at the same speed as the clock), this is the way it appears: the light takes exactly one second to hit the mirror and come back. But suppose we are standing still, watching the clock move by to our right. From our perspective, the clock flashes its beam of light when it is to our left; it hits the mirror right in front of us, and it returns to the clock to our right. From our perspective, the light has gone a longer distance than merely from the clock to the mirror and back; it has also moved to our right. And since the speed of light is constant, going a longer distance means taking more time. So from our perspective, it takes longer for a clock moving relative to us to tick than it takes for a clock at rest relative to us to tick. By our calculations, the moving clock takes more than one second to tick: time for that clock slows down. Yet, the person travelling with the clock does not see the clock in movement. He sees his clock tick at exactly one second. From any static perspective, based on objectively correct calculations, a clock in motion ticks more slowly than one's own motionless clock. And of course, it is not just a matter of clocks. Everything that occurs while in motion, measured from the standpoint of someone at rest, moves slower. If a person is moving along with that clock, time for that person appears to us to slow down too, and that person ages more slowly than we do.

Time dilation is, strangely, symmetrical. When observer A says 'my clock ticks in one second and B's takes two seconds to tick', does B say 'A's clock ticks in half a second'? No, he says 'A's clock ticks in two seconds'. Each person thinks, 'I am the one at rest, it is the other guy who is moving; it is the other guy's clock that is slow'. And in a sense this is right. Inside a rocket ship, we are in a stable inertial environment; if there is milk in the rocket's fridge, it is not flying around. The milk is at rest, like us. So from our standpoint, we send our beam of clock light and it takes one second to return, but when the other guy standing steady on a planet that is moving very fast relative to us sends his beam of clock light to a mirror passing near to us, we can see it must have taken two seconds for it to get back to him. And likewise, he sees from his standpoint that our beam of light must have taken two seconds to his clock's one second.

This generates the twin paradox. One person leaves earth and comes back. The person standing on earth thinks '*I* am at rest, the other person has been moving, therefore the other person's time is slower, therefore when she gets back she will have aged less than me'. The person on the rocket says 'I have been at rest [add some science fiction here to make this sound plausible], it is the woman on the planet who has been moving, so when I get back she is going to be younger than me'. This looks like a contradiction. From each perspective, it looks like the other person should be younger, but it cannot really be both. This is solved not in the special theory of relativity, but in Einstein's later general theory of relativity, which takes gravity and acceleration into account. The person who left accelerated at first but decelerated upon return, and the general theory of relativity says that the time dilations of acceleration and deceleration balance out. So by the time the other person comes back, they balance out to be the same age as the person who never left, which happily avoids the twin paradox (at least on some interpretations).

Almost every aspect of time is overthrown with the problematization of simultaneity. If the present is defined as everything that is simultaneous now, and simultaneity is relative, then the present is relative. If the present is relative, some events in the present are, from another frame of reference, already past or still in the future. Therefore past, present, and future are all relative. It is supposedly unproblematic to distinguish and calculate past, present, and future within one frame of reference at a time. But the cosmological result seems closer to Aristotle's theory that time is measurement than Husserl's theory that time is the lived fulfilment of anticipations. This need not imply that real things and events are relative, only that things and events endure over relativistic time. There is some dispute over whether it is measurement, or time itself, that is relative—whether relativity is an epistemic or operational phenomenon, or an ontological phenomenon—whether time measurements are possible, impossible, or conventional. In any case, if we cannot identify the simultaneous now with what really exists, the now can exist simultaneously with the past and future.

Hans Reichenbach[4] discusses an interesting variation. If we shine a light from our position at P1 (event-1) and it arrives at a distant position at P2 (event-2), and is then reflected back to us at P1 (event-3), we know that event-3 occurs after both event-1 and event-2. But since we at P1 cannot see event-2 at the moment it happens (but only after the amount of time it takes the light from P2 to reach us again), we cannot directly know whether the time at which event-2 at P2 occurred was exactly midway between the time of event-1 and event-3 as we measured them at P1. We would like to assume that event-2 occurred simultaneous with whatever events occurred at P1 midway between event-1 and event-3, but we cannot know this. We observe experimentally that the time it takes to send light from our position P1 to P2 and back to P1 is constant each time we do so. The overall return-trip speed of light is constant. But since we cannot directly see when the light arrives at P2, no experiment can show whether the speed of light was constant during its entire travels, that is, whether the light travelled at the same speed when travelling from P1 to P2 as it did when returning. So while we know that event-2 occurred after event-1, we cannot know which particular event at P1 *between* event-1 and event-3 occurred before, after, or simultaneous with event-2 at P2. Physically, then, there is no meaning to the assertion that two distant events occur at the same time. Various experiments, real and hypothetical, have attempted to save simultaneity at a distance by refuting this uncertainty. But they all depend on the assumption that space is orthogonal and that space-time is isotropic, which can only be shown if the speed of light is constant, which in turn requires the assumption of simultaneity at a distance. Any attempt to verify simultaneity at a distance is circular.[5]

We might like to say that the observers at P1 and P2 should just synchronize their clocks. Then the observer at P2 could later report to the observer at P1 what time it was when she (the observer at P2) received the light. But while we can synchronize two clocks when they are together in a single frame of reference, once they move apart at different speeds, calculation from the different frames of reference show them ticking at different rates. The observer at P2 could write us an email saying that the light reached her at 2 pm, but how would we know that her clock read '2 pm' at the same time as our clock read '2 pm'? The only way to observe if the other's clock is synchronized with our own is to flash a beam of light at it and wait for that light to reflect back to us; but our calculation of when her clock read '2 pm' depends on knowing that the speed of light is constant, which can only be verified by knowing what time it says on the clock at the other end, so the problem is again circular. We can only judge simultaneity at a distance if we can synchronize clocks, which we can only do if we know that the speed of light is constant, which we can only do if we can synchronize clocks.

This has interesting counter-intuitive, consequences. First, Reichenbach proposes that we call two distant events simultaneous precisely when there is no way to determine the time-order among them. We might have thought that simultaneity refers to a precise point of time whereas succession refers to a less determinate stream, but in fact it is successive events that can be ordered without ambiguity and simultaneous events that lie within an indeterminate range.

Second, we sometimes speak of the now as the paradigm simultaneity, such that everything now exists together at once. But Reichenbach's paradigm of simultaneity is not one instant but the blurring over of many. We have to call two events simultaneous as long as the one at P2 can be matched *either* with this moment at P1 *or* that one *or* that one... The moment the light arrives at P2 is simultaneous not with one particular moment at P1, nor with a definite plurality of moments at P1, but with one-among-a-range of moments at P1. In a different way, Bergson too will identify simultaneity disjunctively. The answer to the question what is simultaneous with what is not a correct disjunct but the disjunction as such.

Third, we might have thought that simultaneous events are causally bound together as part of the same event, but instead, distant simultaneous events are separated in such a way that light cannot move between them fast enough to convey causal influence from one before the other occurs. Reichenbach says that this is part of our intuitive understanding of simultaneous events, namely that they occur so close together in time that neither can be the effect of the other. However, the idea that events are simultaneous when there is no measure of their temporal relation would not have been at all intuitively evident, had we not already read Husserl's phenomenology.

There are several ways to handle the fact that we cannot know whether the velocity of light is the same during the trip out and the return. One is to stipulate by 'convention' that event-2 at P2 occurs simultaneously with the moment at P1 exactly halfway between the time the light leaves P1 and the time it returns to P1. Strictly speaking, it seems better not to make any judgement than to make a judgement by convention, but the question of the 'conventionalism of simultaneity' has generated an enormous literature. In principle, if the speed of light were not constant but still objectively measurable at a distance, simultaneity between P1 and P2 might be synchronized at a moment different from the halfway mark of the round-trip duration, that is, by 'nonstandard signal synchrony', but there is no great evidence that this is the case (Jammer, 221).

Conventionalism arises from clocking distant light velocity. Jammer describes attempts to avoid conventionalism by defining relativity without light and clocks, for example by axioms, or by deriving special relativity from general relativity (focusing on gravity rather than light) (145–7), or by coordinating the geometry of Minkowski diagrams (256–7).

What makes the problem worse is that simultaneity is problematic not only across distant frames of reference. We cannot rely on the simple story that 'local' simultaneity is unproblematic within a single inertial frame of reference, that is, within the 'time of a system' of objects at rest, that is, within 'the set of all simultaneities' in that system (140). For even within a system more or less at rest, we still have to move between two clocks in order to measure two simultaneous events. Einstein's usual way to define a 'short distance' is to lay down a 'rigid rod' so that it lie simultaneously at the two distant points. A rod is 'rigid' if moving one of it simultaneously moves its other end. But since the ends of the rod are at least minimally distant, we cannot test whether the movements at the two ends really do occur simultaneously, since we cannot synchronize clocks at a distance. Nor can we attach a clock to a solid body and say they move simultaneously. Nor can we synchronize one end of the shaft of a clock's hand with its other end. Nor, for the same reason, can we even define a solid body (222).

If the constant velocity of light during its return trip cannot be determined, can we be more certain of the one-way velocity of light? Can we transport a clock from P1 to P2 and be confident that it remains synchronized with a clock that remains at P1? Could we at least transport many different clocks from P1 to P2 at different speeds, and get an average (not absolute, but better than merely conventional) measure of synchrony (in effect, using multiple rhythms to save simultaneity) (247)? No, at near light velocity, we cannot be confident that the clock moving away from us remains synchronized with the clock that remains behind, since the moving clock slows as it travels. But if the clock is transported at an infinitely low velocity (in effect, infinitely delayed), is simultaneity saved? No, since from P1, we cannot measure without relativistic effects whether the velocity of the clock as it arrives at P2 has remained infinitely low (242).

Reichenbach posits another consequence of relativistic simultaneity. If two events that are not simultaneous from one frame of reference are so from another, then two moments in a person's life that are not simultaneous from one frame of reference might be so from another. From one frame of reference, a person will have aged relative to an event, and from another, she will not have. Reichenbach suggests the possibility of temporal loops, where an event passes on, yet remains simultaneous with an earlier moment in time in a different frame of reference. Is this time travel? Perhaps a single person exists in two temporal frames of reference. It is not so much that there would be two of me, as that the time at which I am younger and the time at which I am older (the past and the future) coexist in different frames of reference.

After 1905, philosophers attempted to incorporate the new physics into ontology. Often, this meant appealing to relativity to confirm the metaphysics they already held, whether neo-Kantian, analytic, or phenomenological. Whitehead finds that the simultaneity of events within each person's time frame 'prehends' a totality of internal relations.[6] The phenomenologist Oskar Becker finds that relativistic time-lines extend from embodied egos.[7] Gaston Bachelard finds that simultaneity reveals a synthetic a priori[8], and that relativistic simultaneity undermines 'duration', and wins for him his debate with Bergson.[9] The Catholic philosopher Jacques Maritain finds that Einstein's physics are interesting but his metaphysics are ridiculous.[10]

With the advance of general relativity and quantum physics, problems of simultaneity multiply. The details need an expert eye, but Jammer gives some ideas. Since, in general relativity, all four space-time coordinates in a non-inertial frame interact, it is not clear how temporal simultaneity can be isolated even within a frame of reference (273–4). If the frame of reference is a rotating disk, for example, it is not clear whether simultaneity is transitive (282). In quantum physics, spin might suggest action at a distance, a kind of absolute simultaneity (285, 291). But the measurement of light velocity is complicated by quantum uncertainty regarding the energy measurement of the emitter of a single photon. On the other hand, there might be units smaller than single photons by which to measure smaller time intervals: perhaps the 'chronon', at 10^{-24} second, measures the 'atomicity of time' (290).

In the fallout of relative simultaneity, it is not clear whether ontology should multiply the passage of time or bypass it.

McTaggart's b-series and *totum simul*

One way or another, relativity is a challenge to simultaneity. But if relativity shows that there is no absolute way to distinguish succession from simultaneity, and if quantum physics even shows backward causality[11], then ordered succession is undercut, and this may suggest a new kind of cosmological simultaneity. If different frames of reference have different views on what exists together in the now, and how long it takes for events to occur, then perhaps the big picture is that the time-order of events themselves has little or nothing to do with the passage of now, past, and future, and that time is an abstract structure all there at once. J. M. E. McTaggart's (1927) influential idea that time does not pass through the present is often read as the appropriate philosophy for relativity. It seems to me that relativity is better construed as a theory of unsynchronized frames of reference. But even if McTaggart's project neutralizes aspects of relativity, it has its own plausibility.

McTaggart's *The Nature of Existence*[12] defines two senses of time (Volume 2, 10). The 'A-series' is the passing present: past and future are relative to what is present now. The 'B-series' is a relation of before, during, and after, indifferent to whether events are present, past, or future. Both series are transitive and unidirectional. It is not quite that past-present-future is subjective whereas before-after is objective: if time passes through the present, it passes both for objects and for subjects, whereas if time is a permanent order of events, it is so for both physical and mental events. The difference is rather that the A-series property of 'being present' is not a permanent feature of an event. In contrast, the B-series attributes permanent temporal properties to an event: if one event happens after another, it is forever after it.

McTaggart argues for two theses. First, the B-series presupposes the A-series. Second, the A-series is self-contradictory; the present is spurious, so time qua passing present is not real. It follows that the B-series is not real either. Therefore, no kind of time is real. Many commentators accept McTaggart's second thesis and reject the first; for them, time is real, and the B-series is the correct account of it. (Later in his book, in Hegelian, or perhaps Leibnizian or Bergsonian chapters that few readers attend to, McTaggart offers a third, C-series, in which non-temporal relations in a self, which include one another asymmetrically, may under certain conditions give a legitimate sense to A- and B-series (208–81).)

McTaggart's first thesis is that the before-after series depends on a present-past-future series: before-after implies change, and change implies passage through the present (11–12). If there were no transitory present, and all events were either always present or always subsisting without passing through the present, then before and after would have no meaning. Without the passing present, McTaggart argues, we might still narrate events in order, or number them sequentially, but neither would really be an order of before and after. Russell, he complains, tries to bypass the passing present by saying that if one sentence is true at time-1 and false at time-2, and another sentence is false at time-1 and true at time-2, then the first sentence describes the state of affairs before the state of affairs described by the second sentence. McTaggart replies that the time-references in Russell's sentences beg the question: times can only be designated in relation to a change, therefore in relation to time-passage (14). If McTaggart is right, any serialization of time depends on past, present, and future. As we will see later, this thesis is often challenged, so much so that people sometimes attribute the opposite view to McTaggart.

To prove the second thesis, that the present is spurious, McTaggart gives three arguments. These hang on an odd conception of simultaneity. The first is that the past-present-future relation presupposes a non-passing substrate: 'If anything is to be rightly called past, present, or future, it must be because it is in relation to something else. And this something else to which it is in relation must be something outside the time-series' (19). It is not clear why. McTaggart's idea might be that being present measures an event's relation to a temporal grid that does not change over time, so that the present is really timeless and not present at all. But why must there be an outside in order for time to be ordered? Could time not consist entirely of relations among changing moments?

The answer is found in McTaggart's second argument, which holds that the past-present-future relation is contradictory: 'Past, present, and future are incompatible determinations. Every event must be one or the other, but no event can [consistently] be more than one... But [if past, present, and future are real], every event has them all' (20). That is, any given event would have the attributes of being future, being present, and being past. McTaggart notes the obvious objection, namely that 'these characteristics are only incompatible when they are simultaneous, and there is no contradiction to this in the fact that each term has all of them successively' (21). To counter the obvious and maintain that the present is self-contradictory, McTaggart needs a special conception of simultaneity.

McTaggart constructs a reductio against his opponent's scenario that an event is future at one moment, present at a different moment, and past at yet another moment. McTaggart takes this scenario to mean that if an event is past, it is past *at a moment.* To maintain the non-contradictory character of time, one must say that while an event is present at the present moment, at some future moment it will be past, and at some past moment it was still future (21). The consistency of time is preserved in this scenario only by distinguishing *the event* from *the moment at which it is an event.* Since the *moments* are not simultaneous with each other, the past, present, and future of the event do not overlap inconsistently. The event is 'present at a moment of present time, past at some moment of future time, and future at some moment of past time'. This posits a three-part relation: the event, the event's being past, and the moment at which the event is past. To refute this, McTaggart asks what it means for a moment—not the event, but the moment at which the event takes place—to be past, present, or future.

A moment, McTaggart says, can be a present moment, but the *same moment* will also count as a past moment relative to some future time. The moment *is* present, but *at the same time,* it *is* a future moment relative to a past moment, and *is* the past relative to some future moment. These three descriptions are relational, nevertheless they attribute contradictory temporal properties to the moment. It looked like we could avoid the simultaneity of an event's being past, present, and future at once, by distributing the event's stages into different moments. But each moment in turn exists as past, present, and future, so an event that exists in the present moment therefore exists in a moment that is also past and future. So the event itself is once again present, past, and future simultaneously. It is not just that the present *is* present, *was* future, and *will be* past (which would avoid the contradiction of simultaneous times); it is that the same moment *is* present within the moment that is present, and *is* past within the moment that will be past, and *is* future within the moment that has been past. And if we now try again to avoid this result by saying that while a moment in the present *is* present in the present that is present, a moment in the present *is* present in the past only in the moment that *will* be past in the future, and so on, that is, if we now generate another level of moments in order to avoid the simultaneity of past, present, and future, we merely reproduce the problem at the next level (22). At whatever level we place events into a present moment, that moment will be found at another level in the past and future, and the event will end up with contradictory temporal attributes.

There is more to McTaggart's arguments, but the important thing is the odd way in which he doubles the *event*'s being present, past, or future in relation to the *moment*'s present, past, and future. The event's past, present, or future states are not contradictory in themselves. Nor is it contradictory to say that a moment has past, present, and future states. There is only a contradiction because an event is simultaneous with a moment. Only because of this are the event's past, present, and future states simultaneous with each other.

At first glance, it seems odd, but not too odd, to say that an event is simultaneous with the moment at which it occurs. It is related to the linguistic point (on which McTaggart agrees with Russell) that, 'an assertion that N is present means that it is simultaneous with that assertion' (14). When I say truthfully that 'N is present' then N and my assertion are simultaneous. Similarly, if it is true that N is present, then N and the present moment are simultaneous. But this seemingly harmless distinction between the present event and the present moment creates the problem. For while the event is not in the past, present, and future at once, the present moment is represented in all three at once. The result is that the present event is not after all simultaneous uniquely with the present property of the moment, but is simultaneous in different ways with the past and future properties of the moment as well.

On the one hand, it seems right to distinguish the concrete event and the quantity of time it takes to occur. And it seems right that an event coexists with the time it takes. Even if it is odd to say that it takes as much time for the event to occur as it takes for the time to pass, it still seems right to say that the event overlaps exactly with a certain amount of time, that the event is simultaneous with certain moments in time. Perhaps the consequences of the simultaneity of now-content and now-moment, of event-time and pure-time, are awkward but inevitable.

McTaggart's third argument that the present is contradictory is that the judgement of the 'length' of the present is subjective (28). An event is at best 'simultaneous with a specious present'; it cannot objectively be said to be present. One person's subjectively experienced length of time is not a contradiction. But, McTaggart assumes (too confidently), different people experience the same event as being simultaneous with different lengths of time, therefore judgements about what is present at a certain time inevitably contradict each other. So when we claim to superimpose events onto allegedly objective present time, we in fact spread events over contradictory properties.

It is not surprising that many philosophers try to disentangle the before-after relation from the passing present, in the attempt to save the B-series from the contradictions of the A-series (whether they see themselves as followers or opponents of McTaggart in doing so).

Adolf Grünbaum, for example, wants to make the B-series into a stand-alone, A-series-free, definition of real time.[13] The before-after scheme is not about passing through the present, he argues, but about sequencing. He insists that the absence of time-passage does not imply that all events exist at once like images on a film strip; nor does it imply that the universe is a '*totum simul*' or 'block universe'; nor that all events 'exist *simultaneously*'; nor the 'absurd contemporaneity of all events' (139–40). Even though events do not pass, the before-after time-line renders them measured and ordered, short or long, synchronized or staggered. By definition, indeed, the time-line makes 1917 and 2011 events non-simultaneous, since they occur at different dates, one obviously before the other. Yet as the B-series represents it, the before-after relation is unchanged whether it is affirmed in 1917 or 2011 or 1277 or 3001. Since time *relations* do not change over time, they all hold at the same time. In spite of what Grünbaum wants to say, there does seem to be a sense in this picture in which the times of all events exist at the same time—*totum simul*. And if the date 1917 coexists with the date 2011, it will be tricky to explain why two events dated 1917 are more simultaneous with each other than they are with events dated 2011. As McTaggart says, dating without time passage seems to be merely a way of classifying events, and not really a temporal ordering of before-after.

To be precise, Grünbaum does not altogether deny that time passes through the present; he just argues that time passage is mind-dependent, whereas the before-after sequence is mind-independent (333–4). There is no now in objects (though there is causal, temporal, order among objects, he says). The now is added to the event when a conscious being associates the event's before-after time-order with his experience's before-after time-order. A now emerges when 'M experiences the event at t and is aware at t of having that certain experience simultaneously with an awareness of the fact of having it at all' (340). This is an interesting twist. When Grünbaum defines the passing present as the simultaneity of the event with the experience of the event, it is as if describing the present as a moving simultaneity clears the path for describing the before-after order as motionless non-simultaneity.

If it is true that the present does not pass, it follows that the past does not vanish into oblivion. John Leslie argues for the 'four-dimensional block' of space-time[14]. Contrary to the usual assumption, he claims, everyday experience does not exhibit the passing of time. We might have thought of memory, for example, as an experience of succession, of the passing present vanishing into the past. But memory preserves before-after sequences rather than letting them vanish. Leslie likes the analogy with walking down the road: as we walk, the road behind us seems to vanish, but we know it is still there. By analogy, the past seems to vanish when it is more distant from immediate consciousness, but it is just as possible for time as for space that it is still there when we are not conscious of it. My consciousness yesterday is still being lived now, just as Hegel's consciousness in 1807 is still being lived by him *back there* along the road of time. Times, like mile markers, do not go away just because more times and mile markers are added. We may have a 'feeling' of time passage, but our experience provides as much evidence that events do not pass as that they do, as much evidence for a non-passing 4-dimensional space-time universe as for a 3-dimensional spatial universe passing through time.

Leslie is also a philosopher of religion, who believes that if all time is present at once, then we never vanish. Of course, in one sense we are not eternal, since our existence does not coincide with all of time. One's life has a certain time span, but the piece of time in which we exist does not vanish, so we are always living in it. It is not that 1905 was real when it was present, and less real when past, since there is no past, present, and future, only times. 1905 is simply 1905, like chalk is simply chalk. He thinks that this is good if one has lived a happy life, but if not, it is distressing, since one will be living one's unhappy life for eternity. Leslie judges that if time did vanish, for example, if a good life were to happen and then stop happening, God would have made a wicked universe. The discovery in physics of the block universe of time should give people a stronger moral and religious sense.[15]

Where does this quick survey of simultaneity in and around physics leave us? We cannot simply correlate the A- and B-series with succession and simultaneity, and then study these separately under subjective phenomenology and objective physics. For my method of incorporating into the definition of simultaneity every plausible variant, it is a desideratum to synthesize phenomenological convergences and divergences of time-passage with the theories of nature that find simultaneity to be relative, where frames of reference are calculated independently, and where time-order is to some extent independent of the feeling of the passing present. Whether passage is removed from temporal structure or not, time is still a system of differences relative to other differences. In Bergson's terms, we must consider whether the incorporation of physics into phenomenology leaves us with one or many structures of time.

(b) Delay in, but not Observed by, Consciousness: Cognitive Science

Just as Plato uses planets to discuss simultaneity and desire to discuss delay, so today simultaneity is a core concept in physics, and delay in cognitive science.

Benjamin Libet argued in the 1970s for a strong type of cognitive delay, which he calls backward referral of subjective time perception. Patricia Churchland prefers what she calls the postponement hypothesis, a weaker type of delay than backward reference. Tor Norretranders provides a helpful history of Libet's experiments and follow-ups, citing apparently successful replications.[16] Churchland says they have not been replicated, and that the experiments are flawed. Not having run the experiments myself, I discuss delayed consciousness as an interesting phenomenon that *might* exist.

We feel like we experience events in the world in real time; our experience of physical events feels like it occurs simultaneously with those events, without delay. The problem is that the brain takes up to half a second to process the stimuli by which the events in the world affect the body. There is a delay in processing the stimuli in the brain, but subjective experience does not notice it. In phenomenology, we think of delay in the mind of the anticipating subject. But in Libet's experiments, delay is a physical phenomenon, precisely the part of cognition that we do not experience: our normal perceptual experience fails to notice the delay that has in fact occurred in the brain. Phenomenologically, there is no delay, but materially there is. Consciousness compensates for delays in the brain, so as not to notice them.

Libet stimulates people's skin, asks them what they feel and when, and looks to see what is happening in their brains. He sees something happening in the brains before they report their feelings. As a scientific observer, he sees a certain sequence between the time the stimulus is presented on the skin, the time that the cortex shows electrical activities, then the time the person pushes a button or says 'I feel it'. But the person having the feeling and reporting on it from the first person perspective feels the pinprick and experiences herself saying so immediately and without delay.

Why does it take so long to feel the sensation? And why does it not feel like it takes so long? Here are some Libetian experimental findings[17], summarized by Norretranders:

1. Delayed decision: A person deliberately flexes a finger muscle. She reports her conscious decision[18] 0.2 seconds before she actually flexes the finger. But 0.3 seconds before reporting her decision, that is, a half-second before the act of flexing the finger, there is correlated brain activity in the cerebral cortex, referred to as 'readiness potential' (Norretranders, 214–9). It seems that before the person even thinks about moving her finger, her brain is already working on it. Free will looks like an epiphenomenon based on electrical activity in the brain.[19]
2. Delayed consciousness of physical stimulus: The experience of having one's finger pricked can be caused either by actually pricking the skin on the finger, or by subjecting the cortex to electric shocks. When the skin is pricked, it takes just over 0.5 seconds for the person to report the prick, but the person consciously feels that her sensation occurs simultaneous with the pricking of the skin. (We are not talking about bodily 'reactions' to stimuli on the skin, like jerking one's hand away upon being burned; those reactions are indeed close to immediate, and certainly do not take as long as 0.5 seconds. We react well before we become conscious of doing so. We are talking here rather about processing the information en route to becoming conscious.) When, in contrast, the cortex is stimulated directly, it surprisingly again takes about 0.5 seconds before the

person feels it consciously. One might have expected that it would take extra time for the nerves in the finger to send the information through the nerves to the cortex and that the direct brain stimulation would be experienced before the prick. But the reverse order of conscious experiences occurs (231). The person feels the physically pricked skin slightly earlier than she feels the cortically induced prick-feeling. This is a major puzzle.

3. If something else happens in the cortex with enough intensity, *between* the time that the cortex started to show the effects of the pinprick, and the time that the person would normally become conscious, she never becomes conscious of it at all. It requires the *full* half-second of work in the cortex to become conscious of the pinprick.
4. If the cortex is stimulated in a place corresponding to the right hand, and 0.4 seconds later, the left hand skin is stimulated, what happens is that 0.1 seconds after the latter (therefore 0.5 seconds after the cortex stimulation), the patient feels the *left* hand being touched. Apparently without sufficient time to be processed, the left hand skin stimulation piggybacks on the readiness potential caused by the pseudo-right hand stimulus, and overrides it (234).
5. If the stimulation on the finger is weak, there will be 'evoked potential' in the brain, but without consciousness. This might be called 'subliminal perception' (232). Extremely brief stimulation on the finger will generate 0.5 seconds of brain activity along with consciousness; but brief stimulation of the cortex will not lead to continued brain activity or consciousness. The brain does not generate enough delay to yield consciousness of the event. The delay is so essential that we might say that consciousness *is* the delay in the body whereby it knows itself.
6. If the part of the brain that is stimulated is the thalamus, rather than non-specific systems in the cortex, the timing by which it reaches consciousness is closer to that of the pinprick on the skin than to that of the direct stimulation of the brain (237).
7. Other experiments concern the time (0.03–0.1 seconds) that a ball moving in front of a person's eyes takes to be processed by photoreceptors in the brain. The experience of a moving ball does not feel like it lags behind the ball. Indeed, the subject can point to where the ball really is, not just to where it was 0.1 seconds earlier; she can shoot it accurately with a pistol. In fact, the retina looks slightly ahead of where the ball is. A ganglion cell, part of the neuron, engages *before* the ball gets to that position. The neuron tells the person that the ball is there, before it actually is. The brain has a futural or anticipatory character. The body

> does not just act in the present. It acts on the future that is not there yet, and compensates for delays. There is something odd no matter whether we locate seeing in the eye/brain or in consciousness. If seeing is in the retina, it sees the object too early; if seeing is in consciousness, it sees it too late. Somehow, if seeing works, it occurs in the delay, and in compensation for delay.

The delays involved in synchronizing what happens on the hand, in the brain, and in consciousness depend on something occurring simultaneously with something. The physical processes in the brain begin simultaneously with the brain being stimulated; the nerve processes begin the same moment the skin is stimulated. Conscious experience aims simultaneously at, even if it does not occur physically simultaneously with, the finger being touched. Of course, everything takes time, and it is not surprising that it takes time for an event in the world to reach consciousness. What is odd is that it does not feel like it takes time. As Norretranders says, consciousness 'lags behind, but it does not tell us that. It dupes itself' (220) by a 'useful fraud' (235) or 'blind spot' (240). Norretranders' folk-evolutionary explanation—that the brain takes time to settle on which information out of an unmanageable quantity of data is important enough to be conscious of—seems to me, like most amateur evolutionary explanations, ad hoc.

To be sure, some delay mechanisms improve a person's chances for success. For example, there is a relation between children's IQ and delayed gratification. Four-year-old children are given a marshmallow and told to wait 20 seconds before eating it. Those who can do it show higher IQ's and better social and emotional skills years later. Even this finding seems questionable, though, since waiting requires skills besides the ability to delay. Waiting may require imagining alternative pleasures, so maybe it is really imagination, and not delay, that is correlated with IQ.

In any case, the puzzle remains the mechanism of perceptual delay. As Norretranders puts it: 'How does consciousness remain in "real time"? Clearly, we do not live half a second behind events in the "outside world"' (232).

The commonsense reading of Libet's phenomena is that each stage of the event—in the finger, in the cortex, and in consciousness—occurs at a discrete time, and the problem is why consciousness cannot see itself as a latecomer, that is, why consciousness cannot see that it is not simultaneous with the first stage. The puzzle changes, though, if delay is part of the natural structure of consciousness.

For example, John Eccles' interpretation of the phenomena is that the brain undergoes 'rhythmic variations', some of which favour a given action. Consciousness does not so much become aware of events after a delay, as 'keep track' of events, cleverly making decisions at those moments when it can ride the brain waves (222–3).[20] 'Keeping track', the just-in-time economy of decision-phenomenology, is like the structure of delay itself.

Libet's interpretation is that consciousness experiences the event on the finger 0.5 seconds after the cortex receives it, but 'assigns' (232) its experience to real time, that is, close to the same moment that the cortex received it, that is, simultaneous with the finger being pseudo-stimulated. What Libet calls the 'subjective relocation in time' results in conscious experience being 'projected back in time', '*as if* it occurs when the brain puts out an evoked response' (235).

> There is an automatic *subjective referral of the conscious experience backwards in time* to this time-marker, after the delayed neuronal adequacy at cerebral levels has been achieved. The sensory experience would be 'antedated' from the actual delayed time at which the neuronal state becomes adequate to elicit it; and the experience would appear subjectively to occur with no significant delay from the arrival of the fast projection volley. (Libet, 1979, 202)

Norretranders notes that the projection back in time is correlated with experience being 'projection out into the body' (235), but he does not emphasize this enough. Reference backwards is consciousness's way of referring to its own body. Reversed time-order matches the spatial order of the lived body's already-thereness.

Epistemologically, we need the fraud of backward reference in order to synchronize events correctly. Though every process takes time, usually we can simply register that we processed information a certain amount of time after the event occurred. It is not shameful to know that our experience of a star occurs years after the star looked like that. But we cannot manage our bodies in real time unless we take our consciousness of our fingers to occur simultaneously with what happens on our fingers. This synchronization is fraudulent in the sense that it ignores the work in the cortex in the interim, but of course it is not fraudulent insofar as the intentional content of our consciousness of our fingers correctly represents the world around us. As the lower limbs of statues installed on a hill above the viewer have to be foreshortened in order to appear proportionally undistorted, consciousness has to be delayed and then contracted in order to synchronize with the world.

In sum, whenever consciousness is present, is takes itself to have been present earlier. It is always before its now, delayed in principle, reconstructing the now, synchronizing by shunting its contents back into external temporality.

Patricia Churchland[21] makes two objections to Libet's results. The first involves technical considerations of the comparative timing of the stimuli. If she is right that Libet's protocol is defective, then the story is over, but evaluation of the technical controversy is beyond us here. Churchland's second objection is that even if the data are as Libet says, the 'backwards referral' or 'retroactive dating hypothesis' is exaggerated. Instead, Churchland proposes a 'postponement hypothesis', 'a less spectacular explanation; to wit, there is a *delay* in the works, such that the cortical sensation is felt *later* than predicted' (Churchland, 177). The sensation is 'put on hold, so to speak, until it can be "admitted" to consciousness'.

There are difficulties in defining 'on hold' when it pertains to neurons, but the scientific literature in other contexts does refer to 'delay neurons'. The neurons of monkeys who have not yet received a promised reward can be monitored to see what they are doing while the monkey is doing nothing else. This is different from the monkey having an image of food, or a desire for food—neurons associated with images and desires may fire as well, but other neurons fire associated with delay alone. There may be more delay neuron activity if the reward is more desirable.[22] Similarly, a human subject may be told to delay completing a task (asked a question and told to wait before giving the answer).[23] Neurons might record their answer as 'working memory' until it is needed. In this way, holding the past is much like preparing the future. At any rate, something physical corresponds to the 'meanwhile' function. Certain neurons may do nothing but work on delay; their job is time lag.

Churchland's idea is simply that it takes time before the cause in the brain has its effect in consciousness. She says this is simpler than retroactive time-assignment, since it puts the cause before the effect rather than trying to synchronize the conscious effect with the physiological cause. Churchland does not object to consciousness having a 'time-sharing algorithm', which redistributes temporally mis-perceived events into their original ordering (178). Her concern is to debunk Eccles' attempts to infer mind-body dualism from the gap between brain and consciousness. On the latter point, she is certainly right. But it is the former point that we are concerned with. It is not enough to say that there is a time lag between body stimulus and conscious effect; the point is that consciousness locates its intentional referent back to the time when the stimulus occurred, which is intentionally earlier than the time when consciousness actually occurs.

Libet's reply to Churchland[24] is that the mere fact that different functions of the body take different amounts of time to carry out (the postponement hypothesis) does not explain why the different ways of stimulating the skin and various parts of the brain, each of which may begin at a different moment and take different amounts of time to accomplish, all generate experiences that refer the external event to the same moment of time (Libet 1981, 193). Without synchronization, with the same event being processed at different times in different places on the body, one would expect 'subjective jitter' (194), with the same external event seeming to occur a few different times in close succession. But this is not our experience. We need to explain the 'rigidly timed signal' of the intended event (194). This seems convincing. However, it assumes that consciousness has one overall experience of a given event, which is either smooth or jittery. Dennett's alternative explanation of Libet's results does without that assumption.

Dennett's theory of 'multiple drafts of consciousness'[25] is that there is no central position in the brain where the various neural events come together, no conscious audience-member who watches a movie inside the brain, and no single time at which correlated neural events are synthesized. A lot of events take place in the brain, and no internal subject sees them all together. There is a lot of input, and plenty of interaction, but no central processor or interpreter. It is true that when a neural event occurs, we can 'work back' to see when its various conditions had to have been met. And it is true that we can regard a neural event as a point of departure and work forward to the earliest time that a different sort of event can make use of it (Dennett 153). But Dennett rejects the idea that all stages in a neural series refer back to a single original timing signal (157). Dennett distinguishes two interpretations of Libet's findings: a strong interpretation that affirms 'backwards *projection*', as if the conscious experience of the original neural event is 'sent backwards in time'; and a moderate 'backwards *referral*', which simply 'reminds' consciousness of what happened earlier in the brain (157–8). The latter is obviously milder, but for Dennett, neither view is necessary, since there is no single moment at which the neural event-complex takes place, no single moment which can be called the upsurge of consciousness, hence no moment at which brain and mind are synchronized, and no need for the moment of consciousness to be projected onto the moment of the neural event. 'There is no moment of absolute time' (161), so 'a judgement of simultaneity [between reception in the brain and conscious perception] need not itself be simultaneously arrived at' (163).

Perception, for example, is not one thing, but, at different times, consists of events in the retina and various places in the brain: there is no one place or time when it occurs, and no time at which it all comes together and the person becomes aware of it (165). Dennett focuses on the temporal spread of the physical events, but we could say the same about the temporal spread of phenomenological consciousness involved in perceiving, judging, deciding on, and reporting a visual content. The temporal spread is irreducible, and in that sense, Dennett is right to emphasize the 'asynchrony' of the brain-consciousness relation (165). Different organs will be registered in the brain at different times, different questions posed of the person will get answered at different times, and events and reports need not be synchronized. The mind is like a crowd scene, with each bit of the brain shouting, 'Do this!' There may be cascade effects, when neurons shout what their neighbours shout, until the body moves. But (as some roboticists find[26]), goal-directed function is more like swarming than like deciding.

However, Dennett remains focused on consciousness as a factual event, dispersed as it is, rather than as an intentional reference that is about something. The key point remains that when we are conscious of something happening on our bodies, we are conscious of it happening while we are conscious, even though phases of our consciousness are distributed over time, some of which are delayed relative to the event in the world. Even once we know that events are spread across time in phases, it remains a curious phenomenon that consciousness arising *after* some phases of the event is nevertheless referred to the same target event with the earlier phases. The simultaneity of consciousness with an event on the skin is not just an illusion or false memory, nor is it just a phase of an ongoing series. It is an experience of a now simultaneous with another now despite coming after it. The key point is that consciousness targets intentional objects and synthesizes temporally phased micro-events into events that share duration. In a sense, phenomenology has always said this: syntheses of consciousness make common intentional targets out of diverse events with different objective temporalities. The problem is not just how consciousness refers backwards to events at previous times; the problem is how consciousness refers to any discrete event-times at all. How does one act of consciousness refer to the same event over different times and still call those event-phases simultaneous or synchronized? Since the event-phases occur at different times, why does consciousness not regard all events as delayed, or as discretely serial and therefore neither simultaneous nor delayed? And in those cases where we know that there is a time-lapse in processing, why does consciousness feel the prick on the hand at the same time as it sees it (or thinks it saw it)?

In short, despite the likely correct view that both physical and conscious events are composites dispersed over time, without a single unit at which the event 'itself' is definitively present, there is still an issue of synchronization for those conscious events that intentionally target temporally diverse physical events together. In fact, there is a rich theory of synchronization available within cognitive science.

The mathematician Stephen Strogatz's book *Sync: The Emerging Science of Spontaneous Order*[27] discusses mathematical similarities (noticed in many different sciences) across a wide range of synchronized oscillators, from the way that groups of fireflies fall into synchronized lighting patterns, to the self-synchronization of pendulum clocks when attached to the same beam, to circadian clocks that govern sleep patterns.

Among other things, applied mathematicians want to explain the 'delayed sleep phase syndrome': why insomniacs live by a shorter or longer waking day than twenty-four hours. Synchronization normally takes place at three levels: among chemical and electrical rhythms within each organ, between organs, and between the body and the world ('entrainment', 72). Subjects placed in caves for months at a time lose their entrainment with the cycles of day and night in the actual world and develop idiosyncratic sleep cycles. Plotting the curves of their cycles proved difficult until several different rhythms were correlated: sleep cycles, along with rhythms of hormone fluctuation, digestion, alertness, and especially body temperature. (The low period of the cycle of body temperature is closely tied to the likelihood of falling asleep, but these two rhythms can fall out of synchronization, 73, 76.) During sleep, there are relatively independent sub-rhythms like that of REM sleep, and all the sleep-related rhythms are further synchronized with the secretion of the brain hormone melatonin, as well as with rhythms of short-term memory. Rhythms of adrenal gland and liver cells are observed even outside the body in dishes (86–7). It may be that there is a single localized place in the hypothalamus (thousands of oscillating neurons in the two clusters of the suprachiasmatic nuclei) that controls all of these synchronizations (which may go against Dennett's denial of central synchronizers), and there may be a plurality of different 'clock genes' (88–9). This is an ongoing field of research, but the complexity as well as the biological and mathematical successes in unravelling elements of circadian rhythms lead Strogatz to say, 'we might not have a clock; we might *be* a clock' (87).

The literature in almost every science contains fascinating reading on delay structures, from the way that the increasing population of a bee colony creates 'queuing delays' in the performance of individual bee functions[28] to the earliest delay-management in conveying information over the internet (in packets, with built-in redundancy to avoid degradation, trading off acceptable delay and acceptable data loss, with time allowance for encryption and decoding, sent along non-parallel pathways to avoid interception, designed to link up again in real time upon reception)[29] to the psychological advantage of 'coming from behind'[30] in competitive bicycling. Amateur pianists like myself find it more difficult to compensate for losing the rhythm when delays are fine rather than large. Languages that put verbs at the end of sentences delay different elements of interpretation than those with other syntactical order. Each system has its own structures of delay, compensation, and catching up, and each science needs its own categories to capture them.

Domains of science exhibit inexhaustible structures of simultaneity and delay. There is no one temporal structure that founds the rest and no one model of simultaneity and delay that explains natural sequencing. Yet, the most diverse phenomena can be revealed through their simultaneity and delay structures. These are not marginal features of certain objects' temporality but essential to the causal functioning of organic and other oscillating physical systems.

With this backstory about relative simultaneity in physics and oscillating delay in mind, we turn back to Bergson.

Chapter 6

Simultaneity and Delay in the Era of Simultaneity: Bergson

Theme of the chapter: Simultaneity mobilizes virtual and actual forces, purely temporal and quasi-spatial forces, as well as subjective and objective forces to constitute the many-levelled rhythms of memory. The disjunctions of simultaneity and the oscillations of delay lead to the problem of individuating events in time.

I begin with Bergson's *Duration and Simultaneity* for the implications of relativistic simultaneity. Then I move to *Matter and Memory* for the convection currents of delay.

(a) Disjunctive Simultaneity as Virtual Substitution: *Duration and Simultaneity*[1]

Bergson differentiates duration and simultaneity at three levels, based on a distinction between time experienced internally in consciousness and time in the physical and spatial world. Bergson argues that we start with time in experience; second, we transfer the time of inner experience onto our experience of the physical, spatial world; third, we let physical time 'ricochet' back onto internal experience. We end up with three kinds of experience of time: immediate experience of internal time-consciousness, the experience of time in the spatial world, and then a map that refers physical time back onto our experience. The third is then a new kind of internal time-consciousness. Each level of time has its own brand of simultaneity.

Bergson's first principle is that in inner experience, 'time is multiplicity without divisibility. It is a succession without separation' (44). Time is at first identical with the continuity, flow, or passage of inner life. The discrete thing and its states are artificial snapshots of the flow. Fluid duration, the indivisible continuity of past and present, is the true form of actual experience.

Bergson's evidence is that when we look at inner experience, we do not find points of division, but when we look at the spatial, physical world, we do. Both claims are controversial, but my primary concern is the definitions of simultaneity that emerge in inner experience and physical space as well as the kind of simultaneity that emerges when the latter ricochets back onto the former.

One of the perennial issues in the philosophy of time is whether time and space are parallel in structure. We often picture a spatial series as one thing existing beside another and a temporal succession as one thing after another. On a spatial field, we can erect a borderline at any point we like. We might think of time the same way. Bergson's thesis is that time is radically different from space in this regard. One argument is that space is measured by the intervals between the starting point and the endpoint of a thing, whereas in the consciousness of time, which is where time is, one cannot mark the exact instant one fell in love, started enjoying breakfast, or started having an idea. There certainly are transitions over time (from bachelor to husband, student to teacher), so in a sense there are divisions, but these involve qualitative change more than precise quantitative divisions. Temporal transitions begin and end with periods of unclarity. In Bergson's terms, consciousness is a continued continuity. Another argument is that in space, we can take a measuring stick from here and put it over there, thereby measuring whether this and that are the same length. There is nothing similar to do in time. We cannot wait for three seconds of consciousness to elapse and then wait a couple of hours and lay those three seconds of consciousness onto the next three seconds to see if they are the same. The different pieces of time do not exist at the same time.

This is a clever argument. It is extremely difficult for a beginner learning piano to play five notes in the space of four. You play the four notes and try to remember how long it took, so you can then play five notes in the same amount of time. But as soon as you go to play the five notes, you have forgotten. You cannot keep those four notes somewhere in your consciousness—there is no place for them to go—so you cannot superimpose the past time onto your present time. Of course, you can buy a metronome, but this makes Bergson's point. If you can transfer the flow of time to some physical mechanism, the mechanism can beat off a certain number of intervals. But then it is not the consciousness of time that you measure. Obviously, musicians figure out how to hear time signatures without a metronome. But even then, they may be transferring time into the physiology of their muscles, the space of their bodies. They are not literally keeping an amount of time-consciousness for later use.

Bergson's third argument for the immeasurability of time is based on the experiences of different animals (46). It is likely, he says, that most humans have similar feelings of events happening quickly or slowly. This helps explain how time is 'relayed' and 'enlarged' from each person's internal consciousness into a common sense of the duration of the whole world. But other animals likely experience speed differently from humans. Perhaps humans could learn to experience time the way other animals do, for example, to slow down our consciousness and see things that previously were going by too fast for us. But the point is that biological difference shows that time-flow is not species-specific but consciousness-synthetic.

So if time is not divided, why do we think we can divide it up by hours or moments, or that events start and stop? It is because *space is* divisible. Since consciousness acts out in space, and space is measured, we feel that consciousness is measured too. For example, the conscious intent to walk across a room is not divided; it is not as though we have hundreds of different plans, each of which is to pace off the next inch of floor space. But when we walk across the room, we smear our action into space. If we then measure the steps, our consciousness *is* retroactively measured, not as consciousness but as the space covered. We map the steps *back* onto our plan of action and now say in retrospect that when we were thinking about walking across the room, we had calculated the time to do so. We could even say that we spent so many units of time imagining walking across the room. We transfer qualities of space back onto qualities of time, which is possible because we acted out our will in space.

In sum, time consists of flow without divisions. Space consists of divisions without flow. The latter is as important as the former. Bergson's thesis is that when we think about things in space, we see *only* snapshots, momentary states. No matter how closely we put together two states, one after the other, there is no continuity between them; there is always just the one state that there is. It is only in consciousness that we feel the move from one to the next. In a way, space is a huge series of cosmic simultaneities. Even once time is added to it, space-time is still made of all-at-onces. In contrast, duration and consciousness are made up of continuity and never simultaneities.

Having said this, Bergson defines three senses of simultaneity, one of which does apply to consciousness (35). We have seen the first two senses already. First, when we act, we spread ourselves out in space and measure the space. A simultaneous situation in space is referenced back to a single, constructed, point in the flow of our consciousness; hence, we experience that at this point in time, the spatial world all exists simultaneously. We make our consciousness simultaneous with the physical world. We experience our actions in the world at the same time as we experience the world. Bergson calls this 'primary simultaneity' (52). Secondarily, simultaneity can be described without reference to our action at all. Here, simultaneity consists in all that is present in one photographed instant. This is the simultaneity of one state of the physical world with another state of the physical world. It is as if everything in the physical world carried a little clock with it, and we synchronize the clocks.

Third, 'pure time', in its full sense of duration, refers to the immediate flow in consciousness. There is a sense of simultaneity even here, namely, the simultaneity of two conscious beings. This should be interesting, because each internal consciousness was supposed to *not* be divisible into the kind of phases that might be simultaneous. On the internalist definition of time, each act of will has an autonomous time frame incommensurable with other projects. One pure consciousness should not be able to measure its experiences according to those of another. However, once they are both acting in the physical world, they both have the same physical things to measure, so now they *can* both decide that what was in their consciousnesses is measurable relative to the other person. Once there is a physical world, there can indirectly be simultaneities across consciousnesses. Bergson is, after all, not sceptical about whether there is space. We are organisms with muscles and bones, we do move in space, space is observable by us and by others, and when we move in space, we see the other person's body moving beside us.

The ground of simultaneity is acting or moving in space, as opposed to merely perceiving space. If two people merely watch an event together, they might not get the same ideas about what occurs simultaneous with what. But if they interact physically while watching, their joint action provides shared markers for determining that something started at this point and finished at that line. And once different people agree that an event occurred within a given time span, they can synchronize a clock and then move the clock and see if something else took the same amount of time, thereby expanding judgements of simultaneity beyond themselves throughout the world.

Bergson next moves from the divisibility of space to the totality of divisible time. 'Nothing prevents us from assuming that each of us is tracing an uninterrupted motion in space from the beginning to the end of its conscious life' (51). Since we measure time-consciousness by seeing our actions in space, we can think of our whole life as one long walk through space. We complete a journey co-extensive with our conscious life, and our entire history unfolds in measurable time. This is the 'ricochet' effect (53). We 're-ascend' to consciousness, measuring all of our consciousness indirectly through spatial change.

We are now ready to read the key paragraph in the text, which connects (a) the ricochet of space onto time, (b) the totality of the divisible and the indivisible, (c) the intersubjectivity of cosmic time, and (d) the disjunctive logic of contemporaneity:

> Are we thinking of such a journey when we speak of an impersonal time? Not entirely, for we live a social and even cosmic life. Quite naturally we substitute any other person's journey for the one we would make, then any interrupted motion that would be contemporaneous with it. I call two flows *contemporaneous* when they are equally one or two for my consciousness, the latter receiving them together as a single flowing as it sees fit to engage in an undivided active attention and on the other hand separating them throughout if it prefers to divide its attention between them. Even doing both at one and the same time if it decides to divide its attention and not cut it in two. I call two instantaneous perceptions *simultaneous* that are apprehended in one and the same mental act, the attention here again being able to make one or two out of them at will. This granted, it is easy to see that it is entirely in our interest to take for the unfolding of time a motion independent of that of our own body. In truth we find it already taken, society has adopted it for us. It is the earth's rotational motion, and if we accept it, if we understand it as time and not just space, it is because a journey of our own body is always virtual in it and could have been for us the unfolding of time. (51)

There are nine points in this passage.

First, simultaneity begins with substitution: 'Quite naturally we substitute any other person's journey for the one we would make'. The simultaneity of one person's experience with another's now takes precedence over the simultaneity of one person's experiences with the space that person traverses. When I put myself through space, I either watch or imagine someone else acting through space. Primary simultaneity is now my simultaneity with another person's actions.

Furthermore, the extension of my sense of space into that of a second person is extended uniformly and continuously to any action that could be carried out by anyone or anything, so as to cover 'any interrupted motion that would be contemporaneous with it'. While I am acting, I ask myself, what could someone else do? That gives me something comparable by which to measure the time quality in space. Simultaneity requires that I superimpose another action onto what I actually do. Our experience of simultaneity is not an immediate perception of two events happening at once; it is an experience of one event happening, plus the recognition of a possible substitution of another event for it. This requires more than imagination; it superimposes the real and the possible.

The second point distinguishes 'simultaneity'—two instantaneous perceptions apprehended at the same time—and 'contemporaneity'—two flows experienced at the same time. On the following page, Bergson calls them both forms of 'simultaneity': 'I will just talk here about the simultaneity of instants, and in other cases I will talk about the simultaneity of flows' (52). Earlier, he said that in the pure flow of consciousness, there is no way to say that two events occur simultaneously. But here he does talk about simultaneous flows of two consciousnesses.

Putting these two points together, namely that simultaneity substitutes a real event with a possible event, and that simultaneity has one relatum within the flow of my experience, and its other relatum in another person's consciousness, Bergson now makes this third, odd and difficult, point: 'I call two flows contemporaneous when they are equally one or two for my consciousness'.

I emphasize this phrase. Events are simultaneous when they are '*equally one or two*'. 'For simultaneous instants too, what makes them simultaneous is that our attention is able to make one or two out of them at will'.

What is the role of the disjunctive 'or'? Why does Bergson not simply say that events are simultaneous if *two* events—one *and* another—occur at once? Why say 'one *or* two'? Or at least, why would he not say that my consciousness is simultaneous with the world if my unified consciousness *and* the world's multiple states exist at the same time? For that matter, why does Bergson not say that my consciousness is simultaneous with the world if the event is 'one *and* two' at the same time? Why is simultaneity 'one *or* two' instead of 'one *and* two'? This speaks to the *possibility* of substituting one consciousness for another, rather than the actuality of doing so. My consciousness cannot in fact belong to another person, yet it can only grasp its own actions if it allows another consciousness's acts to substitute for mine. This is how an '*or*' exists in the world, how disjunction has metaphysical status. Conjunction in the real world is not hard to think about. But for Bergson, disjunction is the structure of time. Time means that when this event is happening, it *could also have* been possible that another event happen. For X to exist in time means: 'This or some X'. What happens in fact is *not* that other possible event, it is *this* actual one, but it is nevertheless *this or that* (the inference rule sometimes called 'Adjunction' makes this valid) and that is what makes it temporal. The structure of 'not necessarily so', of contingency, is necessary for time-flow in general. But in particular, the 'or' makes simultaneity possible, because it is not just that one thing could happen this time and maybe next time something else. It is that *this* is happening, and the alternative *even now* would be something else, so this is happening at the same time that something else might have been happening.

This is why it is better to say that two things are simultaneous because they are one *or* two, than because they are one *and* two. 'One *and* two' would either refer to two individuals extracted from time in order to be in one photo; or else it would posit a dualism of consciousness and world: as if one thing in space plus one thing not in space equal two things in time. The static ontology of events ignores the flow in an obvious way, but the dualist ontology of events ignores it even more deeply, since by treating consciousness and world as two separate things, it separates the time of consciousness from the space of the world, and entails that simultaneous snapshots in space do not have anything to do with the durations of consciousness that supposedly coexist in time with them. The 'and'-form of actual simultaneous events cannot ground simultaneity; it can only juxtapose two substances that do not have the same time to exist in. In contrast, the 'or'-form of virtual simultaneity allows space to be an alternative medium for a conscious event even though no part of the event is both mind and space. To say how this works, we have to look to the role of will.

The fourth point in the passage is that for two events to be contemporaneous, they must not simply be one or two; they must be one or two *for* my consciousness. It is my attention, or will, that can make one or two of them. There is a choice each time we determine whether two events are simultaneous or not. We choose our actions in space, and we choose which portions of the flow of events to pick out, and those choices determine whether we find simultaneity in experience or not. This is consistent with Bergson's idea that we can speed up or slow down experience. If we walk to the end of the room and glance at what is happening on TV, we might say *Iron Chef* came on at the moment we got to the sofa. But if we focus more quickly at just the right time, we might realize that the commercial was still on when we got there. For that matter, we have to be willing to judge ourselves by the standards of another consciousness, or by the standards of the physical world, to substitute one vision of ourselves with another. To see an action as 'one or two' means we have to be willing to see ourselves either as a purely conscious being or as a physical being. There is an authentic act of will in self-interpretation only if we are willing to think of ourselves in both ways. We have to be vacillating dualists: to think that there is a difference between mind and body, time and space, but to think both are true not in conjunction but in disjunction. Only if we are willing to make this substitution in our sense of ourselves can we experience the simultaneity of our consciousness with the world.

The logic of 'one or two', the will to alternate flowing and static interpretations of ourselves, is needed to define 'flow'. Two events belong to a flow if interpretation can will them equally as difference or as identity, smooth or striated. Similarly, 'possibility' assumes that alternate realities, that is, disjunctions, are real constituents of physical reality. If reality consisted entirely in the succession of one actual thing and then another actual thing, there would be no concrete possibility, only reality. A real possibility cannot by definition be actual, but it also cannot merely be the product of imagination withdrawing from reality. It is the point of differing: not the point of difference between two things, but differ*ing*, where there is no answer to the question whether it is one thing or two. My consciousness might be the same as my action, or it might not; time might exist in the real world or it might not. It is only in this situation that mind and body, time and space-time, are contemporaneous.[2]

The fifth point concerns the numbers in 'one or two'. The differences between a succession, an instantaneous simultaneity, and a flow across which simultaneities are distributed, are reflected in different combinations of ones and twos. We might think of several things present at once as a set. When we add consciousness to a series of facts, increasing the set by one element, flow becomes contemporaneity. Alternatively, we might think of consciousness as an infinite set, and a state of the world as a subset of it. Perhaps the subject who defines the set is not a member of the subset. Following Badiou, set theory might tell us something about how simultaneities work.

The odd sixth point is that we '*could have been*' physically in time before being consciously in time. Bergson says not only that we *can* superimpose consciousness onto physical motion in the form of time, but also that we *have done* it. Not only have we done it individually; we have done it as a society, with shared calendars. What makes us willing to accept time based on the physical world, even though such a science violates our original experience, is that a journey of our body is always virtual in space and '*could have been* for us the unfolding of time'. Bergson's conditional differentiates between saying the spatial world *was* for us the origin of our experience of time, which it was not, and saying that space *could have been* for us the unfolding of time. Contrary to fact statements make time in the spatial world possible. We were not in space-time but could have been. And it is not just that we *can* walk through space, and that space *can* ground our sense of time. We *always* could have.

One of two simultaneous actions is always determinate and measurable, while the other is in flux, and simultaneity involves the superimposition of one action from each of those two series, alternating between two interpretations of time and space.

The seventh point is that time is always added on to space as a difference. 'At exactly at the moment of our passing from the unfolding to the unfolded it would have been necessary to endow space with an extra dimension' (59). The extra, the difference, is what allows us to say that two things are happening at the same time.

If, as Hegel suggests, a spatial point extends into a line, time is added onto space at the moment when the second spatial dimension is added onto the first spatial dimension. When line then turns into surface, time is added onto space a second time. However many spatial dimensions are added, time is the plus-one dimension, the dimension that simultaneously holds events together for a multidimensional being. Space maps the here and there of static representations. If adding time simply mapped before and after, events would remain static. Time-flow is only added when we experience the ongoing present of before and after together, not as tenseless properties but as the subject's capacity to relive her life at many levels (see *Matter and Memory* below). In fact, experiencing lifetime at different levels is an eminently practical affair. Once we know that other species live by different rhythms, we can experiment with disengaging our own sense of rhythm from the normal pulse of our heartbeats, for example, with an artificial heart or through meditation.

So it is not just in the abstract that time is the dimension added to space. To exist additionally in time is to control acceleration and deceleration. Free will is largely the capacity to change rhythms of expectation in time, to control intervals. 'Science counts simultaneities but neglects intervals'. It neglects the imagination that allows my rhythm to become another's, the substitution of different velocities that adds contemporaneity to simultaneity. Time adds its extra dimension because it adds the biorhythmic duration of another life of consciousness into my own.

The eighth point is that adding multiple biorhythms makes time flow into 'a history that has been piled up onto a canvas' (59). Time results in 'history congealed'. Time is fluid but at each moment we can measure what is congealed in it; the process of painting is visible on the canvas. But at each stage in the process, the canvas is a launching point for a next movement. Movement is defined as advance beyond simultaneity. And to constitute now the next simultaneity is once again to add the temporal dimension. At the beginning, it seemed that simultaneity were a notion drawn from the false worldview that bases time on the image of space. But now we find that simultaneity, as the launching pad of movement, is a key notion within the durational theory of time as well, for the reason why time adds a dimension onto simultaneity is that simultaneity was a result of piling up and congealing multiple biorhythms already in process. Time is not just continuous rhythm, like the ticking of a clock, or like the repetition of physical events before and after. Time-flow is both the piling up of and the subsequent differentiation of one rhythm and the next. The simultaneities proper to flow, or the simultaneity of a flow and the world, where consciousness is simultaneous with something, depend on constantly changing rhythms. Non-identity of time-flow constitutes simultaneity. It is easy to imagine simultaneity as two clocks on the same rhythm, telling us which sets of events occur at the same time. Laying static photographs on top of one another, or flipping pairs of photographs on the same rhythm, no doubt produces that kind of simultaneity. But Bergsonian simultaneity depends on switching rhythms, so that consciousness can add its own congealing simultaneously with the world's.

The ninth and final point is that the measure of time in space-time faces 'delays' and 'hesitations' (63). Sometimes we want to get something started, but it is not getting going yet. Or we hesitate. Our starting and stopping-points reflect the obstacle of space in time. Even the best contemporaneities are delayed. When one event is perceived right here, its simultaneous counterparts elsewhere in space will not be experienced until later. We act right now, but space makes us hesitate. For each action, some obstacle prevents it, in its own distinct rhythm, from happening all at once; each action is both indivisible and yet unfinished.

Simultaneity in inner time should not simply reflect spatial simultaneity; it should reflect action overcoming spatial obstacles. In inner experience, we cannot actually perceive simultaneously two things in different places in space; we can only perceive simultaneity virtually, by perceiving one event actually occurring and another event that belongs at the same time in principle, namely by seeing one event acted out and the other delayed. Strictly speaking, since actions are indivisible, we should not say that a phase of one event occurs simultaneous with a phase of another, as though one person arrives at a certain point in a brushstroke simultaneous with another person arriving at a certain point in a dispute with a policeman. The most we should say is that at some points, an activity has been piled up on a canvas and is ready for a new brushstroke, or a rejoinder. In practice, as we saw in Husserl, when two actions are taking place, most of the time when one of them arrives at a measurable stopping place, the other is still in flux. It is the delays and hesitations in inner experience that train us to locate natural stopping-points in the physical world. We know what sorts of spatial obstacles get in the way of movement. Time delays, and hence the quasi-measurements of inner time, are ricochet effects off temporal obstacles. Husserl interprets simultaneity as a variant of succession whereas Bergson interprets the variations of succession as reactions to piled up contemporaneities. For each action happening now, some obstacle prevents it from happening all at once. Each action is taking too long; it takes two different amounts of time at once: it is delayed.

Obviously, Bergson's account could be challenged. Even his first principle, that inner time is indivisible, may not be true. Still, the problem of simultaneity across differentials is a genuine one. How can a simultaneity that is only ever virtual actually be experienced? How do we experience the simultaneity of a virtual simultaneity and an actual simultaneity? How can an actual time be a virtual time for another actual time? How can there be delayed virtual coexistence of actual times? What would have to be added to make the virtual event actually simultaneous with the actual event or to make actual time run on multiple rhythms? How can there be a virtual, but delayed, coexistence of actual times? Can we experiment with pure time and the multiplicity of rhythms to discover and create new kinds of simultaneity in actual perception? When do such moments of congealing actually occur, and what do they sound like?

In Chapter 9, we will explore musical simultaneity-delay hybrids. But first, we need to consider the problem of simultaneity for which Bergson wrote *Duration and Simultaneity*, namely, whether relativity implies that there are many times.

Let us review which kinds of simultaneity Einstein relativizes. Within a given frame of reference, time can be sliced up, and two events can be judged simultaneous without relativity. But time across the frames of reference cannot be sliced into moments of universe-wide simultaneity. Distant events are neither objectively simultaneous nor non-simultaneous for all frames of reference. Bergson accepts relativity physics.

But multiple frames of reference could mean one of two things. (a) It could mean that there *are* multiple, incommensurable time-systems; that there are independent time-series, each with its own way of calculating simultaneity-slices; that time in one frame of reference has no correlate in another. Bergson thinks this is a common but incorrect construal of relativity. For Bergson, if a clock in one frame of reference reads a certain time, there surely ought to be some sense, no matter how extenuated, in which a clock in a different frame of reference reads some other time *at the same time*. (b) Alternatively, relativity could mean that there *is* one overall time-system, with transformation equations to predict what happens in incommensurable frames of reference. For Einstein, all frames of reference use the same physical and mathematical laws, so a mathematician in one frame of reference (e.g. the mathematician inside the train) should know exactly what the mathematician in the other frame of reference (on the platform) judges to be simultaneous. They make different judgements, but they know what judgement the other makes, and why. For Bergson, this means that there is one time-system across the universe even though it is only within each frame of reference that two specific events are simultaneous. Bergson thinks that the very thesis that frames of reference make different, but coordinated simultaneity judgements, implies meta-temporality, albeit without unique simultaneity. Of course, if simultaneity across frames of reference meant that clocks in the two frames would give the same reading, obviously that will not happen. But just as for phenomenology, the existence of a real object entails that observers in different positions perceive different sides of it, not the same side, so for Bergson, the existence of a real event at a singular time entails that different observers in different frames of reference will measure it at different clock times, not the same clock time. Bergson concludes that even in relativity physics, there is one time perceived in many ways, not many times.

Bergson's argument in Chapter 4 of *Duration and Simultaneity*, 'Concerning the Plurality of Times', hangs on the distinction between time experienced as simultaneity and time measured as simultaneity. The former he calls 'real time'. The latter, the measure of distant simultaneity, which involves the speed of light and the synchronization of clocks, he calls 'attributed simultaneity' or 'conventional simultaneity'.

Bergson's first premise, consistent with relativity, is that even when other people's time seems to us to move slower than ours, *we* always experience time coherently. The clock of a person at rest ticks in one second by her experience. Likewise, the clock of the person in motion ticks in one second by *her* experience. Our separate experiences of time are exactly the same as each other's. The difference is that when one person 'imagines' time from the other's perspective, and calculates it on the other's behalf, the person at rest says 'it took your clock two seconds to tick one'. This is Bergson's paradox: the two people both experience and measure time as one second, but each one thinks the other's clock is moving at a distorted speed. Bergson concludes that the plurality of times comes into play only when one person calculates *for someone else.* As long as each of the two people calculates time for herself, time is homogeneous across the two frames of reference. The people experiencing time in two different frames of reference are not experiencing anything different from each other. Therefore, there exists only one time.

Bergson is right that two people moving at different speeds have the same inner experiences of time. But some of his examples seem fishy.

Bergson asks us to imagine two frames of reference, moving at different speeds. In the world that is around both of them, events A, B, C, D, and E take place. The observers in the two frames of reference see the same events taking place, though they see them going by at different speeds. Now, when they are both watching the same event A, both observers in some sense have to see it simultaneously, since A itself only happens once. The difference is that when one person calculates for the other person how much time went by between events, this person *imagines* that the events happened slowly *for that person*, and vice versa. Primary, real simultaneity is what people experience, and that is universally the same, Bergson says. Secondary, imagined simultaneity is only what one *attributes* to the other person moving at a different speed. The reason we calculate other people's movements differently is that the speed of light is constant. But the idea that time is slower for someone else is an artefact of the calculations.

Some of the ways Bergson puts this do not seem right. For example, he says that when two people see an event differently, a third person will be able to see that the events they both experience are the same. But for Einstein, there is no neutral observer to see two frames of reference at once. Perhaps there is something systematically wrong with the attempt to say that the two people's inner experiences are the same. Perhaps all we can know is our own inner experience, just as we can only observe the external world from our own physical frame of reference. Bergson suggests that one can put oneself in the mind of the other person to see that the two experiences are the same. Bergson makes a suspect move when he says that if a person were to see the other person's clock simultaneously with his own, then he would see… etc. The point of relativity is that one cannot in the same instant see one's clock and the other person's. Still, it may be that not all of Bergson's scenarios are undermined by this mistake. There does seem something right about his construal that in relativity theory, frames of reference are not randomly independent, but coordinated. The fact that there are many ways of observing time-flow may still validate a unity of time-flow.

Paradoxically, Bergson says, the reason why some interpreters posit many times is that they think incorrectly that there can only be one true perspective on one real time. Since Einstein proves that there are many real perspectives, they conclude that there must be many real times. Bergson calls the assumption that there is one perspective on real time 'unilateral relativity' (*Duration and Simultaneity*, Chapter 1). It is a natural misunderstanding to think that relativity sees each person moving at a certain rate of speed and assuming 'the way I see time is the way time really is; I have the true perspective on time'. From that starting point, as soon as each person realizes that other people also have true perspectives on time, each person then has to say 'you also have experiences of real time, therefore there must be many times'. The alternative, Bergson says, is to assume not 'unilateral relativity', but 'reciprocal relativity'. Assume that there are many perspectives on one real time. Relativity theory is really saying that when two people move at different rates of speed, they see events differently, but each can calculate the way the other sees it, so there are reciprocal ways to see how other people see. Strictly speaking, though each can calculate what an observer at a hypothetical velocity should calculate, once each tries to calculate what velocity the other observer is actually travelling at, her calculations will be her own, not the other's. And strictly speaking, it is not clear how Bergson knows that there is enough initial coordination to guarantee mutual calculation; but relativity seems to assume this as well. In any case, Bergson posits many perspectives on one real time, albeit with many calculations. Beginning (falsely) with one time implies that there are many times; beginning (correctly) with many times implies that there is one.

What does Bergson think real time is? Is it the kind of indivisible duration that exists only within a conscious being's experience, or is it a property of physical space-time? In Chapter 3 of *Duration and Simultaneity*, he says that the one real time is duration, in the sense that everybody's internal consciousness flows indivisibly. But in Chapter 4, he wants duration to cover the physical reality of the relativistic world, the real time of actual events in order, which people moving at different speeds experience as the time of the same events.

From a postmodern standpoint, we might wonder why Bergson thinks it an unhappy result for philosophy if there is a plurality of times. For that matter, Bergson's own model of virtual temporal disjunctions seems suited to a plurality of times with non-coordinated rhythms. Since Bergson's *Matter and Memory* avoids both the simple unity and simple plurality of time, it may be more suited to relativistic simultaneity, even though it was written a decade before Einstein's special relativity. The theory in that text that the past remains virtually simultaneous with the present, suggests a more consistent disjunction of temporal unity and plurality.

(b) Delay as Rhythmic Oscillation: *Matter and Memory*[3]

Chapter 3 of *Matter and Memory* distinguishes 'pure memory' from 'memory-images'. Pure memory is the virtual possibility of recollecting the past in general, a possibility that a person has throughout her life. In contrast, memory-images are actual mental states that exist at particular moments (138–40). Memory has one relation to the past that no longer actually exists; another relation to a past that does exist, in the form of 'surviving' images (41); and a third relation to a past that circulates through the present, which operates as a 'memory cone' (150f.) Temporal relations hang on whether the past ceases to exist. If we are willing to say, with Bergson, that the past has some kind of ontological status, then we can say that in each of these three relations, the past in some sense exists simultaneously with the present. The past cycles through the present by functions of 'oscillation', 'generalization', and 'repetition'. Bergson's theory culminates in the latter, where the whole past is virtually available in ever new ways to the present. But we have to be able to localize particular memories (171), and this will generate a problem of delayed simultaneity.

'Perception' is the immediate presence of objects. A 'memory-image' is a mental representation of something no longer present. A '*pure* memory' is an event we experienced but are no longer aware of; there is no picture of it in our mind at present. But somewhere, the information still exists. It may be recalled, but there is nothing present now to remind us of it.

Perception, memory-image, and pure memory are not just mental states but also material functions. Perception is an interaction with things, not just a spectator activity. To perceive the shapes, colours, and sounds of physical objects, requires the perceiver's bodily movement, action and choices regarding the immediate presence of the physical world. Perception deploys ideas we already have, and so makes use of memory in two ways. The first is habit, where we re-deploy a procedure, picture a sequence of actions in memory-images, and re-enact them in a situation similar to one we experienced in the past. The second is invention. In a novel situation, it would not be useful to re-enact previous images as is, but the flexibility of past experiences lends them power. This is closer to 'pure' memory: not images that used to be present, but resources.

For Bergson, pure memory is what memory really is. The 'memory-image' is not memory itself. For the memory-image is something that takes place in the present, rather than the survival of the past as such. When we take something that happened to us in the past, and make it into an image, make it usable for the present, we turn it into a present interpretation. Of course, this happens all the time; pure memory can be turned into images. But pure memory is something other than the particular image that it can be turned into in the present. A childhood family vacation, for example, involves one's whole self in many ways. It occupies an indivisible duration in Bergson's sense. When we call it up as images, we may remember one slice of time playing with a toy in the backseat of the car, or another sub-event meeting the cousins. But pure memory itself, before we turn it into image-slices, is not the content of any present experience. Pure memory is the flow that survives in itself, rather than the instrumental image. Indeed, pure memory is not something we could experience. It is the virtual existence of the past. It can be turned into an actual image, but is not itself actual.

Strictly speaking, while the availability of pure memory means it exists simultaneously with present perception, the two forms of engagement with events are not on a commensurable time scale. Memories are virtually available, but do not exactly exist in actual time. To be sure, the date of that family vacation is located at a past moment in time. And the actual past event obviously occurred at a time that is straightforwardly not simultaneous with the present, earlier on the same time-line as the present actual event. Still, the memory is now a virtual effect. Just because it is not on the same kind of time-line as the present event, it can be simultaneous with the present. If there were only one kind of time-line, past and present obviously could not be simultaneous; but if there are two kinds of time-orders (one being a time-line—a series of presents; the other something like a time-storehouse—the pure past), then the two can coexist virtually simultaneously (though obviously not in the same sense in which two events at the same present moment on a time-line are actually simultaneous). In short, the memory of the past still exists in the present, albeit not in the way the actual is present. Memory persists in a quasi-temporal form of being. Like arithmetical propositions, memories are available procedures not limited to any actual time, past or present.

Nevertheless, each time it is activated in a person's bodily perception, the virtual memory is made simultaneous with a present. Bergson calls the body 'the place of passage' or the 'hyphen' (151). The body exists in the present, and is the site of present perception. For that matter, the bodily process of perception includes sensation, which registers the way the world looked in the immediate past, as well as motor activity, which is directed towards the future. So at each actual present, bodily perceptions generates a specific time-line of actual (not virtual) past, present, and future.

That perception registers past states of the world follows from Bergson's theory of sensation (*Matter and Memory*, Chapter 1). In a causal system, each material object has butterfly effects on its neighbour, which in turn relays those effects, so that each object has continuous spillover effects. Wind causes ripples on a lake, which leads to a fish eating a speck of algae, which leads to ... etc. As Bergson puts it, when physical objects move, they vibrate and give off 'images' of themselves, which the next object translates into its own form of behaviour, which then reflects images of other objects. Bergson is both an extreme materialist and an extreme idealist. When a person speaks, for example, sound vibrations propagate through the air, and an ear receives an image. Literally speaking, the sound image is not constructed out of the mind of the hearer; the image is made up of the voice of the speaker. Similarly, a visual image consists of light bouncing off an object, splitting into different colours as it moves through a medium, and ultimately being received by an eye. Again, the image is a part of the thing seen, not first and foremost in the seer. The object's movement, reflected in an organism, constitutes perception. We could call this either a materialist theory of images or an image-theory of matter.

Of course, the image has to hit a sense organ and be interpreted. But insofar as the image comes from something else, sensations are direct presentations of the past. Sensation perceives the world as it was before, connecting the perceiver to the stabilized past. Perceiving means living literally in the past. And since vibrations relay effects forever, the past never dies out. It is past because it is an image already in circulation, but the past never falls into the past, it remains present; it is just a question of where and how it will circulate to a given receptor. On the other hand, as we receive and perceive, we also engage with objects, and the body in motion gives perception an impending, futural quality. The things we act on confront us literally with the future. For Bergson, sensory motor activity is thus the meeting point of perception and will, and thereby the meeting and prolongation of past and future. By virtue of having a body, we confront past and future 'at the same time', now. 'Having extension in space, my body experiences sensations and *at the same time* executes movements' (138). When an animal avoids an attack, for example, its pain is in the past, and its evasive manoeuvre is in the future. If the animal hits a beehive with a stick, it coordinates two objects in space, prolonging their simultaneity until the activity is complete. Since perception is clearly simultaneous with action, the past is simultaneous with the future, in the present. Anything with a body, with muscle-memory and the will to self-preservation, is a past-future hybrid.

The body thus connects past and future in particular activities. In pure memory, by contrast, the past survives en masse.

Bergson's thesis, covering both body memory and pure memory, is that 'there will no longer be any more reason to say that the past effaces itself as soon as perceived than there is to suppose that material objects cease to exist when we cease to perceive them' (142). It is obvious that when we cease to perceive an object, it may still exist. Bergson, like block universe advocates, wants to prove that the past continues to exist even after we stop perceiving it. He also wants to prove that consciousness directly perceives the past, qua past. Just as perception apprehends an image that comes from the object itself, so the direct content of consciousness *of* the past *is* the past.

Bergson begins with two spatial analogies. First, just as there exist parts of space not present to us at any given time, so parts of time that are not present may still exist (142). Second, just as space extends outwards from each object, so we can draw a 'diagram of our near future' (144). However, analogies with space are not decisive.

Bergson says that the commonsense reasons why some people think that the past does not coexist with the present, are weak (145–6). One is that we cannot re-eat this morning's breakfast, because it is no longer in existence. Bergson admits that past objects cannot be used in the same way as present objects but responds that the fact that they can be used at all shows that they exist.

Another reason why people think the past ceases to exist is that remembered pasts are unordered, and unordered things cannot exist. It is certainly true that we can remember today's breakfast either before or after remembering that childhood vacation. But the inference assumes that everything that exists has an ordered place, and that a thing we can recall whenever we like with could only be a fantasy. Bergson responds first that even though we can retrieve past events in any order, they do retain order in the form of their dates of occurrence. Second, the past is useful precisely because it can be taken out of order. In contrast with space, the relative disorder of past time is what makes it real. It is helpful for a bricoleur of memories to find them in a clutter, so he can use them for purposes other than their original ones. They are most useful when their use is underdetermined.

Bergson's best argument for the persistence of the past is phenomenological and applies more to pure memory than to body memory. The past is given as an object of two very different kinds of everyday memory experiences. One kind of memory (closer to body memory) is ordered. We want to remember where we left our keys, so we deliberately retrace our steps in imagination one by one, remembering each sub-event (powering past the inevitable gaps) in the order in which they originally occurred. It is as if each event is dated down to the second, and memory rolls back the calendar (whether by starting at the first image in the chain and remembering forward or by starting at the end and remembering backwards). We may not know which memory-image we need, so we run through them all until we find something useful.

The second kind of memory (closer to pure, virtual memory) does not depend on order. A past love flashes into our mind out of the blue, or we concentrate on recalling a melody without the context in which we originally heard it. We may call up a recent or distant past event out of order; it may arise because it is useful for the present, or it may just as easily bubble up by chance. If we find it interesting to survey a wide range of memory-images, we can do that, since they are all potentially available at any time.

The first kind of memory experience produces memory-images in succession; the second accesses pure memory all at once. The two kinds of memory represent two different forms of time, and clearly both are real phenomena of experience. Both are genuinely temporal, highlighting duration but incorporating a role for divisibility. Neither succession nor coexistence is the fundamental structure of time; time has two, independent structures. In the context of simultaneously accessible pure memory, the succession of images constitutes a delay of complete access; in the context of successive memory-images, the simultaneous clutter of possibilities constitutes a delay in selecting a usefully particular memory. For the romantics, events never get going, since each step is deferred, and the beginning of the past is delayed; for Bergson, nothing ever goes away, and the ending of the past is delayed.

The two types of memory also represent two types of tensed ontological status for events. A single event has two ongoing onto-temporal statuses at the same time. As a process, it is actually taking place; as a set of possibilities, it is already being registered as a reusable paradigm. The first is its present status; the second is its past status, which lets the event be engaged in the future. Even while it has the first, present, status, it simultaneously gets the second, past, status. In one sense, an ongoing event is present now, but in another sense, it is already virtual and so already past. It is a body, and already an image, and already a resource for producing new images, passing away yet already preserved. Deleuze will exploit this phenomenon.

Pure memory could be an image again. We are tempted to ask where all these memories are stored (148). It is easy to say that images are located in the brain. But Bergson's theory is that the idea of storage, whether in the brain or in consciousness, assumes that memories are actually somewhere in the present, whereas his point is that memories are only potentially in the present. Pure memories are not actual images, but virtual possibilities, and not anything that can exist in the present. The past can only be 'stored', as it were, *in* the past, and does not exist anywhere in the present, except if we say that the past qua past exists in the present. Retrieving a memory is not like accessing something stored in the present but like returning to the past where the memories are and only afterwards creating a new mental image in the present. The possibility of memory depends on the continued existence of the past; nothing in the actual present explains memory.

So to explain what and where memories are, we need a diagram not of a present physical thing, like a brain, but a diagram that puts virtual and actual things onto the same picture. This is what Bergson's diagram of the memory cone is supposed to illustrate: virtual existence coming into actual existence and fading back into virtual existence (152), virtualities and actualities cycling through each other. Imagine a cone with a base at the top and a point at the bottom. The broad base of the cone, AB, tapers off to a point S at the bottom, touching a plane P. The plane P represents the totality of the present, of which the point S is my present experience. The base AB represents the original sum of my recollections of the past accumulated in memory. The cone can be sliced horizontally at any point from the top of the cone to the bottom, at A1B1, A2B2, etc., representing different moments of past experience. The more memories I accumulate, the longer the cone extends, as the point S of the present moves farther from the first plane AB. The cone represents an overall volume of memories, to which new present growths add. P is always a new point, and as it moves, new planes become distinguishable in the cone. This vocabulary of time passage is not ideal, particularly on the question of volume. Some features of the diagram are not clear, like whether AB represents our first infant memories, or instead represents the broadest range of all our memories. But its purpose is to show that memory material is transmitted from the heights of pure memory down to the precise present point P where the action is (153).

Sometimes, the actualization of pure memories is blocked. After strokes, amnesia and aphasia leave people without memories or language skills to express them. Bergson's view is that we never lose pure memory, only the capacity to make a memory-image out of it. It is as if an amnesiac forgets how to retrace his path between rooms of a house, and thus how to test the simultaneous existence of objects past and present.

So how does something virtual, which exists in the past, descend into the present and become actual? If the past is preserved into the present, and living through the present means living on the tip of the past, we are left with odd questions like, when am I living? When is now?

Bergson describes three functions by which the virtual past circulates through the present. The first is 'oscillation' (162). Metaphorically speaking, the memory cone vibrates and items slide through the interstices, the way a treat filters down through a gumball machine to the chute, or water percolates through coffee grinds, or convection currents fan through an oven. The process is not necessarily controlled or deliberate. We do not know all the memories we have (if we did, they would be images), and we cannot always access them at will. Something virtually in the past, but not an image, has to be shaken out to become an image. A lot of random access memory material drops into consciousness that is not helpful. We may obsess on it uselessly, or simply decline to pay attention. Though inefficient, a simple oscillation mechanism would probably generate enough memory material for an organism to live by.

But memories can also get actualized in the present by a second function, namely generalization: 'resemblances', 'universals', 'simplifications' and applications of ideas, organize past events into structures relevant for the present (156). If each memory were a specific image of what happened in the past, and did not ricochet off other ideas, it would come into present actuality in its original form, and would never seem similar or useful to any different experience. Images cannot explain generalization (as Berkeley says). On the other extreme, if memories were retained only in the form of universal concepts, that is, if we retained general conclusions from past experiences but nothing of their specificity, then in a different way we would again not be able to apply them to specifics in the present. Generalization requires both that a specific memory be flexible and that once flexible, it be applicable, or we might say, rigidible. It must lose its original specificity and regain a different specificity by moving through the memory cone, circulating through other memories while we are not paying attention. Each memory is transformed as it works its way up into the present. In their trajectories, past perceptions are as dynamic as present experience. Just as we solve problems while we sleep, when dreams loosen them up and melt them into other ideas, thinking when not thinking, so memory puts an idea into a collective and makes it interact with other ideas. Ideas cannot help but expand into collections, just as collections contract into singular ideas. Images circulate through a fluid medium, which tosses globules to the surface and pulls them back down into the mix. It is not quite that memories grow organically in our unconscious, or that they have acorn-like potential to become oaks (and only oaks), or are fated to shape us in later life. Some remain dormant, others are funny. Expansion and contraction are the rules for Bergson's (more Hegelian than Humean) associationism: resemblance, continuity, and contiguity are inherent in idea-possession. In short, memory is not containment, but the reconfiguration of perception in variations, combinations, and conceptual generalizations. Its 'ricochet effects' produce a force of mutual ideational attraction.

In sum, if the memory cone were simply an oscillator, perceptions would fall back into the present unchanged, simultaneous with the present but not suiting it, as if they were shards of another life dropping into the present. In contrast, if the memory cone is a ricochet attractor, perceptions are transfigured by the time they are called into the present, which integrates into the simultaneity of the present. Even so, the ricochet model feeds us the past only a few relevancies at a time. How is it that we feel our whole past life is contained in our memory? How does the memory cone allow the entire past to coexist with each present?

The third function suggests that at each present moment, the whole cone is pressed down onto the present (169), like a Bodum coffee maker. A person's memory interprets her whole past. Her present re-enacts every past experience in some way or other. Childhood experiences are at play in current professional life, career choices are at play in breakfast choices. Of course, some pasts have more salience here or there; some contract with tension, as a block, in a given present, while others remain relaxed. Or to change the metaphor, some past moments oscillate so slowly that we do not notice them at a given time or move too quickly in and out of relevance.

> Everything happens, then, as though our recollections were repeated an infinite number of times in these many possible reductions of our past life... they thus enter into an unlimited number of different 'systematizations' (169).
>
> If, here again, we imagine a number of possible repetitions of the totality of our memories, each of these copies of our past life must be supposed to be cut up, in its own way, into definite parts, and the cutting up is not the same when we pass from one copy to another (170).
>
> There are always some dominant memories, shining points round which others form a vague nebulosity (171).

The 'domination' metaphor suggests that memories struggle to control the past. Gadamer likes to quote a passage from Aristotle's *Posterior Analytics*[4] that describes how concepts work. Imagine an army being routed. The soldiers are running confused in different directions. Now one soldier decides to turn and face the enemy again, thinking: this might not be so bad; we may be able to stand up and fight. Another soldier sees her and turns too. Suddenly the whole army is facing the enemy and working together. Aristotle says this is what the mind does when it makes a universal concept. Too often, idea fragments rush in all directions, in a state of panic. Once in a while, a thought-fragment stands its ground, and memories join it in the struggle for conceptual coherence.

Though they do not appear to be present at the same time, every memory is ready to be repeated at every moment in the present. Each is relayed virtually at each moment, and each, or all, might be actualized at any given moment, albeit in a different way relative to each new present. Every moment of the past is simultaneous with the present, though only occasionally do they all coalesce into a figure, a direction or a meaning. In Leibnizian terms, we might say that each memory (or in Freudian terms, each symptom) expresses the whole of a life from its own perspective, though each with its own degree of intensity. Each momentary experience is a re-copied version of our past life in a slightly different form.

A theory of memory ought to explain both how one's whole life is remembered, as well as how particular pasts can be picked out at particular present moments. It is fine that Bergson imagines a single, univocal memory cone, rather than many. If the cone were split into different regions, individual memories would tend towards either suppression or obsession. Nevertheless, there have to be routes, or search engines, that run through the cone as determinate connectors. Individuating memory contents, after all, is a condition for innovation. So even after describing all three functions of the memory cone—oscillation or juxtaposition, generalization or interpretation, and repetition or life-memory—we still need to say how a single specific memory can play a larger role in the present than others. But the more we explain how memories interact, the harder it is to explain specific interactions. If all of my memory is present, why does it not all get smashed up into one? How are coexisting events distanced from the present? When I drive a car, how do I follow the highway rather than reminisce about childhood toys? For that matter, if memories circulate in a convection oven, how could I keep straight which memories were formed simultaneously with which others, that is, which events happened when I was twenty years old and which when I was thirty? How do I place a certain collection of memories onto the same AB plane? How do I preserve my sense of having once been a child? Is it merely a myth that we remember slices of our own time-lines? Do memories interfere with each other to such an extent that we do not remember particular events at all, but make one big simultaneous narrative out of all of them?

Bergson raises the topic of 'localization', but he does more to generate the problem than to solve it.

> The process of localizing a recollection in the past cannot at all consist, as has been said, in plunging into the mass of our memories, as into a bag, to draw out memories... By what happy chance is it that we just hit upon a growing number of intercalary recollections? The work of localization consists, in reality, in a growing effort of *expansion*, by which the memory, always present in its entirety to itself, spreads out its recollections into an ever wider surface and so ends by distinguishing, in what was till then a confused mass, the remembrance which could not find its proper place. (171)

The danger is that Bergson begins with so much simultaneity that memories not only coexist but coalesce, and that as a result, he cannot account for local simultaneities or distinguish events that occurred simultaneously at one point of time from those that were simultaneous at another time. Blending the virtual memory field with actual perception is necessary if there is to be any memory at all. But to solve the problem of localized memory, we have to conclude that the majority of blending has to be delayed. Universal simultaneity requires distancing pasts from one another by delayed simultaneity.

This is related to the problem in Hegel regarding how a particular temporal line can cut through a temporal map. We might want to say that certain memory lines, or line clusters, are bound together with more intensity, or by stronger themes or decisions or histories of effectiveness. But how are such distinctions possible? And even supposing that memories are encoded with their original dates, and can be accessed by that index, why do all the dates not flood into memory as soon as any one of them does?

Perhaps the problem is not too dire, since we can concede that many memory chains can be picked out, highlighted, and superimposed at the same time, polyrhythmically, out of the memory background. As Bergson emphasizes, events look different when perceived quickly or slowly. The rhythm of an event has to be lined up with the rhythm of perceiving it, otherwise the event either looks like nothing, or like hundreds of other events. If there is a Bergsonian solution to the problem of preserving successive clusters of coexistence, it depends on rhythmically condensing and contracting, accelerating and delaying, simultaneities.

On the other hand, we do not want to solve the problem of localizing times too well. We need coexistence to operate even within local temporality and across temporal distance. Writing a history, after all, posits a transitional point (or several) but also draws multiple lines through it, condenses lines into themes, and connects that history to others at as many simultaneous junction points as thinkable. And for the sake of free personal history, we struggle to draw ourselves into as much world history as we can, re-enacting its planes and summing them up through the present point, all the while keeping our distance from undesirable nodules. Reliving an event at different levels is not the same as calculating time from different frames of reference, but both require event selection under asymmetrical expressions.

To see how particular pasts are foregrounded, or at least recycled in a discernable rhythm, it would help if we rendered simultaneities and delays concretely perceivable, either in an experience of the virtual out of phase with actuality, or in a practical sphere like politics, or a sensuous temporal medium like music. But before moving to concrete regions, I want to show how problems in the great simultaneity philosophers—Husserl and Bergson—might be solved by the great delay philosophers—Derrida and Deleuze.

Chapter 7

Simultaneity and Delay in the Era of Delay

Theme of the chapter: Delay is embedded in every now, so simultaneity occurs at a distance from itself. If Derrida is right, delay precedes succession, and Husserlian problems can be solved. If Deleuze is right, past and present coexist, and Bergsonian problems can be solved.

I have presented Husserl and Bergson as the early twentieth-century Simultaneity bookend of the problems of simultaneity and delay—with Husserl coordinating simultaneities in order to structure anticipatory delays in succession and Bergson coordinating delays in order to structure the virtual simultaneity of memory. I now present Derrida and Deleuze as the late twentieth-century Delay bookend. The unresolved problems in Husserlian phenomenology (staggered time-lines and retained delays) are resolved within Derridian deferral; the unresolved problem in Bergson's durationology (preserving distance within simultaneity) is resolved within Deleuzian coexistence. Simultaneity has not been damaged beyond repair; nevertheless, delay takes over the conceptual space in which temporal categories are defined.

A book on Simultaneity and Delay could be written solely about Derrida, and another about Deleuze. Of philosophical texts before the Derrida-Deleuze generation, it is fair to say that certain passages deal with simultaneity (like Kant's third Analogy of Experience), while others deal with delay (like Plato's *Philebus*). With Derrida's post-phenomenology of temporalization and Deleuze's ontology of pure time, it is no longer possible to separate the two topics. I will touch on Derrida's innovations around différance/deferral and his analyses of simultaneity in Aristotle and Heidegger. But there will not be time to do justice to his discussions of the social delays in gift exchange, in verifying friendship, in the consequences of hospitality, in messianic waiting, or in regard to textual delays: the grammatological delay of signification, the literary contretemps of Romeo and Juliet, or the metaliterary critique of the idea of 'early and late' texts.[1] Likewise, Deleuze's complex theory of time is entirely about simultaneity and the temporal distances it contains, and the whole Deleuze/Guattarian edifice of abstract machines expressing themselves in machinic assemblages hangs on aionic simultaneities and their chronic distributions. Because the field is too broad, and because in the case of Deleuze I have discussed these themes elsewhere, I focus in this chapter on the interventions of Derrida and Deleuze in the problems left behind by Husserl and Bergson. In the next chapter, I touch on political simultaneities and delays in Derrida and Deleuze.

(a) Delay as *Différance*: Derrida's Solution to Two Problems of Time in Husserl

Once philosophers (from Hegel onwards) proved that things undermine themselves, aporias may count as solutions. When Derrida shows that the now is the impossible simultaneity of the non-simultaneous and that the instant must first return from a future that will have been its past, this is not merely a challenge to the classical thesis that the now is present when it appears. It is true that Derrida is not simply deciding on a new thesis about the now, since questions about whether the now does or does not have a given property (e.g. singularity, or existence in the present) are by his argument undecidable. But if the best philosophies of time produce aporias, they are time's own real aporias. They are undecidabilities that solve problems by inventing new conceptual topologies. It is not that we do not know what to think about time; it is that the solution to what time is may contain contradictory properties. Both sides of an aporia have to be right, but each makes the other wrong; as soon as we have chosen one side, we realize it is the other side we need to choose.

The problems arising from Husserl, namely that simultaneity is distributed across lines of delay, and that remembered anticipations carry a temporal line other than the line of presents, are not anomalous puzzles but paradigmatic phenomena of temporal difference.

In 'Ousia and Grammé'[2], Derrida asks, with Aristotle, whether a now is a point of sameness or of difference. If the now is sameness in time, it implies coexistence. But coexistence implies being at the same time with something. Either it is coexistent with some other time(s) or it is coexistent with itself. Obviously, if there are many nows, they cannot exist at the same time. Spatial points coexist, but temporal points exist only in succession. Each now is cancelled by the following now. In short, if the now is sameness in time, it is not the same as other nows. At most, it exists at the same time as itself.

That the now exists at the same time as itself seems obvious, but it becomes less obvious once we ask when a now perishes. If the now perishes at the same time as it exists, then it both exists and does not exist at the same time, which is impossible. But if it perishes at a different time than it exists, then (a) it would have to perish at a moment that it did not even exist, which seems impossible, and (b) it would not be able to perish until another now existed too, so that two nows would exist at the same time, which also seems impossible. The perishing of a now requires it to both exist at the same time and not exist at the same time, as itself. As Patrick Miller shows, this paradox problematizes one of the most central Aristotelian axioms, namely that a proposition cannot be both true and not true at the same time.[3]

But if we consider not the coexistence of one now with another, and not the coexistence of a given now with the moment when it passes away, but just the moment of the now insofar as it coexists with itself, is there still a problem? There is, if we consider when the *meaning* of the now takes place. A now is not just a formal place-holder but a moment when some content is present. When is the content or meaning of the now graspable: in the now or in the process in which a series of nows passes by? And when is the coexistence of the now with itself graspable? Derrida arrives again at the aporia.

> Coexistence has meaning only in the unity of a single, same now. This is *meaning*, sense itself, in what unites meaning to presence. (62–3)

The usual idea is that the meaning of coexistence is within a now, not across nows; indeed, the only meaning of anything is within a now. Judgement takes the form: this, at the same time, means that. Meaning is synthesis: two terms that mean something for each other must coexist in time. Having one idea and then later another does not count as one giving meaning to the other. Once traditional philosophy accepts that there are now-points, it cannot help but conclude that something has meaning only if we can see it all at once in a now.

In short, meaning is a synthesis between two things that coexist. But since two nows (and their contents) do not exist at the same time, it follows that two nows cannot have meaning in relation to one another. Yet, if two nows have no meaning in common, we cannot even say that succession is a meaningful connection. On the other hand, if synthetic meaning *is* present at a single moment of time, it makes a simultaneity out of nows, which ought to exist in succession, and ruins the flow of time. Coexistence across nows seems contradictory, but without it, there is no succession at all. How can we consistently articulate the idea that synthesis connects at one time meanings that arise in succession?

If we abandon the thesis that a now can only coexist with itself, the alternative is that meaning-synthesis brings two different nows into contact in the same now. For Husserl, this is not a problem. Husserl holds that each now has retentions and protentions of others in its background that are still, or are already, present. There is not just one autonomous content of the now experience. For Husserl, a now from the past is not happening any more, yet consciousness can go back to it without leaving the present. But for Aristotle, if a now is still present, even as background, then the old now never disappeared and time did not pass. The new now simply got put on top of it; if two nows exist at the same time, they are the same now. This might be the conclusion we have to reach, but it is a direct paradox.

Derrida's conclusion is that once Aristotle asks whether there is one now or many, both alternatives end up impossible. If two nows do not coexist, then there is no synthesis in the now, and nothing present; but if two nows do coexist, then there is no difference between them, nothing to synthesize, and no time to grasp what is there, so again no presence. For Derrida, the solution is to posit not times but delay: not nows but deferrals. Deconstructive delay starts as a critique of the concept of the now, but it ends as a substitute for the now and a solution to the problem of time.

Derrida concludes on behalf of Aristotle that the meaning of the now *means* a contradiction. But the now is the possibility of that impossibility. 'Each now coexists with that which cannot coexist with it'. And every now is the same impossibility as the other nows; it is the same impossibility all the time. 'The impossibility of coexistence can be posited as such only on the basis of a certain coexistence' (63). It is impossible for nows to coexist, but we know this because we put all the nows together. To explain how nows contain coexistences that they cannot be part of, we have to say that the impossibility of coexistence depends on 'a certain *simultaneity* of the nonsimultaneous' (63).

The persistent '*maintenance*' (63) of nowness means that there is a coexistence of the past, present, and future in the now. The present includes non-presence. The now *is* the passing into the future that it will be. The now includes in itself the time that is not now. The now is somewhat not now. This is something like a process theory of time, but process is just as peculiar as the non-present present: the non-occurring stage in the process is what makes the process possible, precisely by keeping the present moment off-balance. In a sense, it is the pure presence of the now (rather than its particular content) that anticipates the future, but it is a presence that exists by impossibly being the same as the presence of past and future.

The coexistence of meaning within a now would be impossible without those nows that do not coexist. The propositions 'everything is simultaneous' and 'nothing is simultaneous' entail the same paradox. It is tempting to conclude that Derrida has no theory of time, since there is no border between one moment of time and another, and thus no border between temporal and non-temporal structures. Perhaps, Derrida says, the whole idea of 'being at the same time' is not a temporal notion. Maybe it is structure, meaning, being-with, or synthesis, and not time, that makes the coexistent and the non-coexistent possible at the same time. Maybe the *in*temporal event that creates time-synthesis, which does not take place at any given time, makes sense of time by non-temporal means. The Greek *hama*, 'at the same time', may be an intemporal adverb, affirming temporal coexistence only to undermine it (64). Maybe the apparent succession of nows is really more like an eternal present, and maybe the flow of time is all anticipated at once. One way or another, we have to represent at the same time content that is not in the same time. This is why it might be better to say that time-consciousness is built out of delays and not out of nows at all.

Speech and Phenomena[4] discusses simultaneity in consciousness. Consciousness must, in the presence of a single now, experience the meaning of an event that covers a series of nows. A meaning experience has to be present, Derrida says, citing Husserl, 'in the same blink of an eye' (67f). But we have to see (as meaning) what we do not see (the percept of some other time). Derrida's conceit is that part of seeing is the blink, the time when the eye shuts. Since any experience includes not only an object but also the eyelid moving up or down—or the process of interpretation, or of writing—no actual experience is present all at once. In Husserl, one instant is already onto the next, returning to the past from the perspective of an ideal interpretation which it has not yet completed. The future does not exist, but the now returns from the future perfect, it has already been being its future; it is what it will have been. As Derrida puts it, the now repeats the future. The present is the back end of the past as well as the front end of the future. Indeed, each direction is reversed as soon as it is described. The unspoken past means something unpredictable, which the future re-enacts. The recent past is contained in the now precisely as the future will recall it, as the now's future recall. The future is the next re-enactment of the past as delayed reaction. This is why Derrida says that 'the return is a bending back that is older than presence' (68). Bending the eye back, looking back meaningfully for the second time on what was seen, is a pre-condition for looking immediately the first time. Of course, Husserl agrees that we anticipate what we are going to see. But the implication Husserl did not anticipate is that we never see something in the now; we always see it in the now afterwards. Part of what must be in the now is still in some other now. So simultaneity *is* always two different nows maintained at once.

In one way, this situation 'brings into question phenomenology itself' (66, 74), but in another way, it is the essence of phenomenology to combine the principle that meaning comes together in the present now (63), with the principle that there is a flow of consciousness.

Derrida's '*Différance*' essay[5] takes up the differential now. The verb 'to differ' differs from itself by having two senses. On the one hand, difference means distinction or discernability. On the other hand, it expresses the interposition of delay ('detour, *délai*, *retard*, reserve, representation', 8). In the first case, two things are distinct, but each is self-same, and the difference is stable. In the second case, the thing in question does not exist when it is in question. When it arises, it differs from expectations. Deferral implies content different from itself. In order to make a distinction, we have to defer whichever side we are thinking of at present, while we consider the other side first. So deferral (the second difference) is the condition for distinction (the first difference).

Derrida's literary-linguistic argument is that the source of meaning of a given text is prior to any enunciation; the meaning is broader in scope than any now attributes to it. Therefore, its explication is postponed beyond the interpretation written at any given moment. The meaning of a text is already retained before the first retention in the first attempt to interpret it, and therefore deferred beyond its last anticipation, to the point where the classical notion of 'meaning' itself is subverted.

Derrida's temporal strategy is articulated across the '*Différance*' essay. Every few pages, Derrida says, 'let us start at the beginning again'. After a few pages explaining why he spells the word '*différance*' in a special way, a new paragraph begins: 'for the same reason I do not know where to begin…' (6). Then, after writing about space and time in language: 'let us begin with the problems of signs and writing since we are already in the midst of it' (9). It is commonsense to say we begin where we are, in the middle of things. What is striking is that we begin with what does not begin. Then, after discussing Saussure on the arbitrary origins of words, a paragraph begins, 'let us begin again' (13), this time with the topic of temporal beginning, using Hegel. Finally, near the end, Derrida says, 'the history of being begins with the forgetting of being' (25). Inside a quotation from Heidegger, there is another beginning, Being before beings, affirmation before language.

Language affirms a system of relays, enunciations are re-iterations, and responsiveness is the delayed effect of the past's desire. Here the essay ends.

For Husserl, the present incorporates the recent past and the immanent future while it is still now. Even though the past and future are not happening now, we run through them within the present now. The now contains time-particles that are not now. The durational now is presence, but includes non-presence. In Deleuze, this leads to the puzzling question: what makes the now pass into the next now? Derrida's quasi-Husserlian answer is that the now's own non-presence *is* in presence. It is not that some moments are present and some not, and that together they make up time. It is that the non-present is prior to, and the condition, within the present, of the present.

Yet Husserl rejects Freud's notion of 'delayed reaction', or *Nachträglichkeit* (*Lectures on Time-Consciousness*, Appendix IX; Derrida discusses this rejection in *Speech and Phenomena*, 71 and '*Différance*', 21), experience that comes to awareness only after a delay. For phenomenology, if we are not conscious of something now, we cannot save it for later and be conscious of it then. A later experience can only remember what is happening now if it retains the shadowed-off presence of what is actually experienced now. There is no presentation of a world that no longer exists. But while Husserl cannot accept the unconscious, where content is repressed, stored, and experienced after a later psychic development allows it to emerge, he must accept that preconscious perceptions below the threshold of attention can be examined later. But even to accept the latter, phenomenology has to affirm multiple protentions in a single perception. It need not accept a realm of being that we have no consciousness of, but as Daniel Giovannangeli puts it, phenomenology accepts that consciousness arrives after reality already exists (it is not phenomenalism, after all). While consciousness is originary presence, it is delayed relative to being.[6] The theological-turn phenomenologist Jean-Louis Chrétien holds that 'every inauguration has some delay, and even an initial consciousness already has a past'; every experience has a history, just as every self already knows its responsibility for others. For Chrétien, deep in the ontological past is an 'immemorial truth', an unhoped-for, 'latent gift', of God.[7] We do not need to pursue this last point. But phenomenology cannot avoid dealing with what, for consciousness, precedes consciousness.

Quentin Meillassoux argues that this 'ancestrality' problem, the fact that there was a world before consciousness entered it, ruins phenomenology's claim to first philosophy[8], but in fact, this strong form of delay is part of

Husserl's theory of presence via anticipation and fulfilment. Intuitions always fulfil judgements about what was already there. The stuff that pre-exists consciousness is, for phenomenology, precisely what future experience makes a delayed intentional reference to. Without the wedge of delayed temporality, the now that passed out of the future would remain static. The durational now that runs back and forth outside the instant would be stuck in the now. If we think of the now only as a short duration, we might imagine that the distant past and future are unstuck from the now. But fortunately for anticipation and fulfilment, once time runs a little bit into the future and past, there is no limit to how far it will run. Distant futures and pasts will all be 'inscribed in' the present (*Speech and Phenomena*, 75). Still, delayed content needs a way for the not now to be *not too* included in the now. But this again is the problem: if time is inscribed within the now rather than passing away from it, Husserlian time will be meaningful but not thoroughly temporal.

Indeed, for Derrida, time cannot be continuous duration. Time requires a pre-included future that is neither merely included in the now, nor merely differentiated from it. The now must preserve other moments of time without either inclusion or exclusion. The past and future must be present in time but not before, during, or after the now, that is, present in time but as a hiatus, a temporary deferral of time itself.

There are four implicit theses on delay in '*Différence*'. First, grasping a definiendum is postponed until one achieves the definiens (8). Second, time is like economics without reserve (20).[9] Third, each present is *re*-presented in a memory, an anniversary, a cyclic renewal, a traumatic repetition, a paradigm, a cautionary tale, a spin-off, or an archive. Fourth, delayed desire suspends the will. There are thus four characteristics of delay: postponement, reserve, re-presentation, and suspension.

I focus on the second thesis on delay. Whereas the classical temporal structure of delay distributes nows equally throughout continuous time (as if living in the now is always good), Derrida proposes to squander time, like the principle of untrammelled pleasure of situationist politics of the 1960s, to put all of the now off and yet want it again, to have everything twice. Reckoning the distribution of time, like gift giving, requires delay in the exploitation of a reserve. When we give a gift, we do not want the same back immediately, nor do we accept nothing in return; at some time later, we will want the past back with interest. In one sense, gift and exchange happen at the same time, but as always, Derrida asks, 'what does at the same time (*en même temps*) mean?'[10] The fulfilment is to take place at a later time, but delay is not just a sequence where one thing happens after another. It is a property of the present in relation to what may come. Delay is the opposite of putting off into the future the events that belong in the future; instead, it lives by the principle of the unresolved future without waiting for it. The consumer of delay is not the subject of enlightened self-interest whose self-denial defers gratification. The one who seeks pleasure in delay wants it all now, yet will affirm the same now's subsistence again later.

The traditional idea of holding the future in reserve preserves the old distinction between now and not-now. As long as the not-now is outside the now, but will come next, an ethics of patience and moderation follows. Do what is appropriate now, and then later do what is appropriate then. Do not spend all your savings now, because you will have more nows to spend them in later. Alternatively, if the not-now is inside the now, then by the time you get to something, you will already have done it. Do not wait for the future in order to do what is in the future, for by the time the future arrives, the event will already have been done in future perfect, and you will not have the chance to enjoy it. So you had better be doing the future now. Squander your efforts now, and then later, squander them again. Temporality without reserve squanders its entire duration in each now and yet repeats the entire duration again in its not-now; it fulfils its entire anticipation immediately and yet withholds that anticipation for fulfilment again later.

Delay constitutes a new kind of expansive present, not by experiencing a process gradually as it becomes available, but by consuming what is unavailable. Delay succeeds in holding time together, where the model of anticipation and retention, with its unsustainable presumption of the continuous now, fails. But if delay assembles past, present, and future, without the usual ordering principle, what kind of time is this? As the 'Blink of an Eye' chapter of *Speech and Phenomena* (76–7) ends by asking, what is time-consciousness if it is not grounded in the now, if we are conscious not now but only later, if self-consciousness is not real-time 'auto-affection', if the time I experience myself now is later than now?

This produces a final twist. As long as we think of the now as present in the now, it looks like Derrida's philosophy is all about delay. For if more of an event is not happening than is, then the event even as it takes place is as much delayed as it is current. But since the now is not self-enclosed, the future which is not in the now ought also to be performed and lived in the now. In this sense, a philosophy in which the now is deferred into the future is not a philosophy of delay after all, but of unreserved temporal presence. After all, affirming the not now in the now does not mean that part of the now is held off for the future. Derridian delay does not imply that we should wait until the now's other half arrives. The now is never going to arrive in that sense. The time to do what cannot yet be done is now.

Let us review the two problems of time in Husserl (see Chapter 2) and Derrida's resources for handling them. (1) The problem of simultaneity began with the breakdown of a single now into many object-directed time-lines. Divergent stopping-points have to be synchronized. The solution requires a paradoxically staggered simultaneity. (2) The problem of the retention of anticipation was that the past's future has to be forgotten in order to remember the present's past. The solution requires a paradoxical delay of the already.

The first problem was that the same now must pass at several rates of speed at once. Derrida's solution is that since each now is inscribed in other nows, they contain its rhythms as well as their own. Each now is part of as many connecting lines as it has not nows as its parts. This generates impossibilities in the time-line, in that the same now is happening at many different times, and the same stretch of time contains both a small and a large number of moments. But since the now repeats its future by anticipating it, this impossibility is the very nature of time. This is why the now passes at multiple speeds. Derrida can solve a problem that Husserl could not, since Derrida accepts the radical interpretation of the shared premises that the solution requires.

The second problem in Husserl was to explain how memory content retains its anticipatory character even after those anticipations have been fulfilled. Derrida's solution is found in his theory of delay. Texts and experiences do not conceal interpretations beyond our grasp. But each reading is an occasion of affirmation, both of the time that passes from one reading to the next, and of the reading that changes purely due to the passage of time, of what 'remains to be read' ('*Différance*', 25). An experiential content never entirely gets fulfilled, so its anticipatory character is not abolished even when it is remembered later. Even when the experience of asking someone on a date is recalled alongside the historical answer, the question still 'remains to be read', and one can still be surprised years later by what one hears in memory. The rough simultaneity of the answer with the question includes the self-differentiation of the delayed reactions of both parties. The question has been fulfilled by its answer, but is also re-presented in another act, where one does not take yes for an answer, where will is suspended by life, where the future prompts a question already dealt with. For Husserl too, there are always some anticipations in an experience that have not yet been fulfilled by the time it is remembered. But that would not explain how memory retains the feeling of anticipation of those aspects that have been fulfilled. Derrida can explain the latter, because unlike Husserl, he accepts the extreme consequence of phenomenological temporal synthesis, namely that each expression affirms all signification everywhere, that each moment is affirmed in a staggered way with other moments at all times, hence that the readability of memory content is affirmed when it is recalled, precisely as transient, undecided, and anticipatory. The fulfilment that would have cancelled anticipation is delayed by time itself.

(b) Simultaneity as Pure Past: Deleuze's Solution to the Problem of Time in Bergson

For Husserl, consciousness flows in dynamic succession, within which simultaneous slices can be imperfectly constructed out of rhythms of delayed convergence. For Bergson, duration is so deeply successive that simultaneous slices falsify it. Yet memory is so dynamic that it allows events to circulate and succeed each other backwards and forwards in time. The result is that events are simultaneous (or contemporaneous) in a new sense. For Bergson, circulatory succession and multi-levelled simultaneity are the same thing.

Deleuze makes Bergson's implications explicit. Simultaneity does not halt the flow of time; it is precisely simultaneity that constitutes flow. Succession without simultaneity would rigidly track habitual responses to proximate conditions; rigid succession excludes flux. In contrast, the simultaneity of pasts, lived at many levels of time at once, allows free movement of interpretation and action. In this simultaneity on a grand scale, there are no rigidly ordered sequences, so assemblages of events form freely. One consequence is that the most dynamic phenomena can appear the least active. The most radical self-reinterpretations may occur in involuntary dreams, or when blurting something out, or when one's resources are forced by delayed reaction into unfamiliar courses of action.

Deleuze turns Bergson's 'pure memory' into 'pure past', or even 'pure time'. There is some justification for translating psychic into ontological terminology.[11] Just as memory has one form that places images in succession and another that preserves images simultaneously, so an event falls under two different temporal categories. An event (a) passes as the ongoing present; the same event, at the same time, (b) is a pattern or diagram of relations that can recur in other events later. The point of a reusable pattern is not that it was actually used at some former present, but that it functions as a pre-existing model, and in that specific sense, functions as the past, for other events. 2 + 2 = 4 as a psychic event passes through someone's mind at a given moment; but the same formula is the same thought anytime anyone thinks it, and even when no one does. It circulates through past, present, and future alike, always having been true. This virtual existence through past, present, and future is what Deleuze calls the pure past. If we were to call it a 'potential' event, that would suggest that it grows naturally out of delimited contexts; by calling it 'virtual', Deleuze suggests a resource available in any context, waiting to become operational. To say that a virtual event exists in the pure past means that it is ready to branch into divergent possible worlds, that it is simultaneously either one or two, that its temporal status is not determinate until a decision operates on it.

For Deleuze, the 'past' is not a set of former presents, but the mode by which events that are not present intervene in events that are. We can say that the pure past circulates through past, present, and future; we can equally say that an event in the pure past is fluid, and is not located in an individuated time. Like Bergson, Deleuze holds that there is no place to store an event, that it can only be, as it were, stored in itself, that it preserves itself, that virtuality is a temporal-ontological status. To be sure, an image of the past can be stored in the brain or in consciousness, but the past as such *is* the preservation of events. On the surface, Bergson's phenomenology is more modest than Deleuze's ontology. But already for Bergson, when we dip into memory, we leap from one kind of reality (the present kind) to another (the past kind). When there is an event in the past, and we think it, our thought literally exists in the past. It is the past that holds our thought in it and shapes its form and content.

This is similar to the story that the past exists back on the road behind us. But the road metaphor suggests that the past exists in a determinate form, that there is a series of images back there to pluck again. For Bergson and Deleuze, the past contains events in their virtuality, not in their actuality. And each time we go into the past to pull out an image, we return to the present differently. At the moment we make it present, it becomes determinate; but until then, the bag of stuff that we reach into is partly indeterminate, since the flow of the event generates disjunctive multiplicity.

Deleuze's most graphic and appealing argument for the existence of the pure past (*Difference and Repetition*[12]) is that without it, the passing present could not be explained. Elsewhere, I have analyzed the many details of this argument.[13] Here I will only sketch the simplest line, emphasizing simultaneity and delay. If present moments were the only kind of time that really existed, the only explanation of why the present passes would be that another present comes along and takes its place—which is absurd. Therefore, since we know that the present does pass, something other than that present, or any upcoming present, has to explain its passage. The passing of the present moment must be explained by something about the pastness of that moment, not by its (or any present's) presentness. Pastness is therefore a temporal structure of its own, by which events become past. In short, it is not that the next present comes along and makes the last present the past. The last present already had a pastness character, independent of its character of being present, even while it was occurring. It has its past character at the same time as it is present. The past is simultaneous with the present. All pasts are simultaneous with every present.

There are many controversial premises and sub-conclusions to this argument. But its appeal is that it is indeed ridiculous to think of a new present somehow forming itself before it becomes actual, then somehow displacing the actual present, transitioning into the latter's position, entering the realm of temporal actuality and pushing its predecessor out into the ontological region of the past. However, the very fact that this is such an absurd picture should make us wonder whether the theory that time is the passing present, has been turned into a straw man. Surely few philosophers ever thought that pre-formed future moments come along and dislodge presents, sending them into the heap of previous presents. Those philosophers (like Husserl) who think that the present passes by its own power, and do not posit an independent 'pure past', nevertheless do not think of the present as a series of static moments, some of which are waiting in the wings. Instead, they think of the present passing as the momentum of intentional objects. As we watch a car race, for example, we do not see a focal point of the present as the cars are whooshing by, which is replaced by another focal point right after; we see a focal duration of the present as each car passing the finish line is followed by the next. The present is an immanent sequence of substitutions and expansions, based on the power of the engines, the project of the race, and other subjective and objective dynamics. The present does not need something added to an instant in order to pass; it is its own logic of passing. Should we conclude (a) that Deleuze intended his story of dislodging presents as an appealing image but not a serious argument against the passing present, or (b) that theories of the passing present really do collapse into an absurd model of future moments with the power to dislodge?

If the model of a new present waiting in the wings seems silly, it nevertheless does seem difficult to explain why a new present arrives, and why the old present has to become past. Why does the old present not just remain present alongside the new? Why does the present not prevent change? Why do future moments not all rush into the same moment? It is as much a problem if the 'next' present is delayed as if it arrives too soon. Once it arrives, there must be no delay in the departure of the last, otherwise there would be a present with no present in it. The presencing of the one and the pastification of the other must occur simultaneously.

Deleuze's strategy for solving the problem of presents dislodged by their successors is to give an entirely different analysis of present, past, and future—not as three stages of time, or three ekstases of one structure of time, but as three different, independent but coexisting structures of time, each with its own ontology and its own definitions of past, present, and future. (1) 'The present' refers not to the present instant but to a process model of time as a whole, where the present explains the past and future: the passing present expands by retaining the implied past and anticipating the implied future. The present is the dynamic succession of actual events. (2) 'The past' refers to a diagrammatic model of time, a set of patterns for possible event-variations that can be instantiated at any time, virtually existing at all times, equally past, present, and future. Not just individual events but series, sequences, and before-after relations have this kind of re-instantiability. Whereas the present contracts and expands into passages with past-present-future stages, the past survives by circulation rather than stages. The past is the coexistent simultaneity of virtual events. (3) 'The future' refers to an eternal recurrence model of time: though events relay patterns of the past and are instantiated in a present, they appear differently on each reliving, and this difference recurs in the event's future. The future is the undetermined, determin*able* character of time, in which events drawn from the past communicate, whether by choice or chance. In the empty future, desire and our capacity to search the past for resources and invent new assemblages are the only limits. The future consists of delayed reactions of simultaneity.

There are, of course, philosophers who think that one of these three models of time describes the whole of time (Husserl, McTaggart, and Nietzsche more or less respectively). Deleuze's theory, which we cannot verify in full here, is that all three describe genuine phenomena, that each depends on the others to explain some of its own features, and that all three must be true at once. Time is a strange, multi-levelled phenomenon, with three independent sets of structures that exist simultaneously. Its levels cannot be synchronized. Present, past, and future are staggered series.

This three-levelled analysis explains, though not on its own terms, the problem of how the present becomes past. Since an event is clearly a memorable pattern while it is still passing, its pastness coexists with its presence. It is just this kind of pastness that we remember afterwards, that survives in its original formal and virtual state into the present. This assemblage of possibilities is what we are influenced by, and can use and continue exploring just as when it was present. This is a plausible account of how it is that an event becomes past by virtue of what was always its pastness, of how the past of an event can be substituted for its presence. It is not quite an account, however, of how one present is replaced by another present. It may account for how a present moment survives as a past, but how does a new present arrive and displace it? It had seemed we were obliged to avoid an accumulation of presents where nothing passes away. But if we have already explained how the past arises, perhaps it no longer matters if we say that the present gets larger and larger and still remains present. After all, the protention-retention process is like a single expanding experience within presence. Perhaps the present does retain images and continue motor perception *in* the now. Deleuze defines 'the present' in just this processual way. The upshot of explaining the past without appeal to the passing present is that the passing present can accumulate without endangering the past.

The structures of presence and pastness generate two distinct senses of simultaneity: correspondences in the ongoing present that expands, and patterns in the past that survive throughout the passing present. Both also generate senses of delay: the present in its unfulfilled anticipations, and the past in the disjuncts between memorable diagrams (or abstract machines) and situational assemblages. There are two simultaneities and two delays (three if we analyzed the future), and the two pairs of temporal structures coexist, irreducibly simultaneously, in each event, which makes each event a staggered delay.

If time really has these levels, they should be visible in experience. When Heidegger's broken hammer broke open the living present, hit the pause button, and revealed a level of being without anticipations and fulfilments, the danger was that it drew that region outside time. Deleuze's pure past is something like that paused region of being, but it is still in time, a region of temporal experience, a new kind of time-field to explore.

Cinema

Because the time forms of past, present, and future coexist, one kind of simultaneity has to be expressed in another kind of time with which it is not synchronized but overlaid. A whole series of pasts has to be experienced in a present, and a single event has to be re-stored in successive presents. There are delays within the expanding present, and between one past and another past distant from it, but the most severe case is reserved for delays between simultaneities past and present. There are two major configurations: the simultaneity of many pasts with one present, and the simultaneity of one past with many presents. The fact that the plural past pivots on a singular present, and the plural present hinges on a singular past, gives us resources for explaining how the total simultaneity of pure time can be narrowed down into a succession of singular events, and how individual selections can be made out of the memory cone totality.

Deleuze[14] finds perceivable illustrations of time-structures in cinema: respectively, coexisting 'sheets of the past' in *Citizen Kane* by Orson Welles, and sequential 'peaks of the present' in *Last Year in Marienbad*, written by Alain Robbe-Grillet and directed by Alain Resnais. Sheets of the past consist of different events of the past superimposed on screen at the same time. Peaks of the present consist of different present moments at which the same past event is played out. Sheets of the past operate when different stages in a life survive into the present; peaks of the present operate when a person relives the same stage in their life in different ways at different times.

Cinematically, sheets of the past can be represented in depth-of-field. The camera films in one shot (using the appropriate lenses, film stock, lighting, length of shot, angles, POV and countershots, montage and sequence) a mansion interior, with the distant background in focus. A character walks from background to foreground. He passes objects surviving from different stages of his life as he moves forward, walking through layers of the past. The distant past, the intermediate past, and the present are piled up in sheets, presented simultaneously to tell the story of tragic succession. This is Welles' version of Bergson's memory cone. Each sheet of the past is a virtual time-world that can be entered and explored further, as when the reporter investigates which sheet of the past contains Rosebud. The opening of *Citizen Kane* advances towards Xanadu in deliberate steps one by one, but by the end of the film, the sheets have been piled up into a single composition. The past is waiting for us to walk through.

The periods of time in a person's life are lined up, but we walk through it diagonally, finding ourselves in it here and there depending often on chance turnings. The delay in understanding our past selves does not always mean that we have too little sense of history; it can mean that destiny is waiting for us in the past. Walking through a scene, the participant becomes his past selves. In *Mr. Arkadin*, the title character hires a detective to report the past he says he has forgotten. Every time the detective gets close to someone who knew Arkadin in the past, that person has just been murdered. Mr. Arkadin, it turns out, is responsible for the murders, but this raises the problem of over-possession by the past: how can one make parts of one's own past *in*-capable of being recalled? How can one avoid having to go back through one's whole past all the time? In Bergsonian terms, if the whole past coexists with the present, how can we engage just one line of our choosing and let the rest remain in delayed animation?

Paradoxically, since the past contains options not limited by the present, the pure past functions much like the future. A film that delves into the past in such a way that the past has not yet happened, comes closest to presenting a pure past time-image. There are, of course, many reverse-time films: *Terminator*rial loops and *Mementos*; straight line backwards *Betrayals*; *Total Recall* technologies; *Back to the Future* romances; traumatic *Irreversibles* seen in reverse; jammed up flashbacks in the *New Rose Hotel*. But too often in films, reverse-time motifs open the past only as predetermining backstories. It is not upsetting when narrative time flashes back into events that already happened—indeed, everything makes sense that way. The crazy thing about time-order is, rather, when time moves forwards, when the event opens an abyss of alternatives and switch-points in the time-line that we see too late. It is not the pasts that actually took place but the pasts that are still delayed and their forgotten futures that open up new lines.

The plurality of pasts piled up on the present need not over-determine the present with destiny, as long as the pasts are as open-ended now as they were when first assembled. For better or worse, an underdetermined past cannot block a time-traveller from revisioning her ancestors. Of course, no one can prevent the past from surviving, but there are different ways to salvage sheet-material from the pile-up.

Deleuze contrasts Welles' depth-of-field sheets of the past with the storytelling peaks of present in *Last Year of Marienbad*. A man approaches a woman and expects her to remember the affair they had at the same mansion a year ago. She seems to have no idea who he is. Sometimes the evidence favours him, sometimes her. Two incompossible time sequences coexist, distributed across different characters. The conceit of the film is not that the present contains two views of the past but that the same past leads simultaneously to two different presents. The past makes room for several possibilities, and each, when actualized, becomes a singular time-line of its own—both narrower than the past it expresses (since it selects only some of the past's possibilities) and broader (since it adds new anticipations and anxieties).

Welles uses time-compression as a primary means of rewriting narratives, particularly in the Mercury Theatre adaptations (with writers John Houseman and Howard Koch) of long works of literature for short radio dramatizations. Slow lead-ups make rapid-fire climaxes intelligible; a rich past can be assembled quickly into a decisive blow. But building a past requires a lot of technique. Only by combining clever abridgement, alternating fast and slow pacing, interjecting radio silence and sound effects, and relying on the audience tendency to sudden panic, was the *War of the Worlds* radio audience in 1938 made to believe that within about twenty real-time minutes, aliens could blast off Mars and arrive on Earth, the worlds' armies could mobilize and lose, and the planet could collapse in chaos.

Time-manipulation distances elements in a simultaneous whole from each other, staggers the selection of simultaneous elements across time, multiplies and hence de-determines the future, and delays the presence of the sempiternal pure past. It solves the problem that Bergson left behind.

What makes time stagger is the simultaneous multiplicity of time-lines. Delay occurs not because a finale is withheld. The delayed finale finally does happen, but it is still delayed because another incompossible finale happens alongside it. From the perspective of one frame of reference, the other is out of joint, and this puts the first out of joint too. Deleuze's philosophy of time is a multi-levelled simultaneity-assemblage, but in its multiplicity, simultaneity is delayed. If we live our lives at different temporal levels at the same time in different ways, each moment of life has its lived time at many times later than the time it is formed. The very structure of simultaneity *is* delay.

It is this delay

Near the end of the chapter on time in *Difference and Repetition*, Deleuze considers the delayed impact of childhood events upon adult events. 'There is no question as to how the childhood event acts only with a delay. It *is* this delay' (163).

> The essential point is the simultaneity... of all the divergent series, the fact that all coexist. From the point of view of the presents..., the series are certainly successive, one 'before' and the other 'after'. It is from this point of view that the second is said to *resemble* the first. However, this no longer applies from the point of view of the... object = x which runs through them...,which establishes communication between [two events] or the forced movement which points beyond them... The question arises

> how to explain the phenomenon of 'delay' (*retard*) that is involved in the time it takes for the supposedly original infantile scene to produce its effect at a distance in an adult scene that resembles it and which we call 'derived'... The two series coexist in an intersubjective unconscious... The childhood event is not one of the two real series but, rather, the dark precursor that establishes communication between the basic series, that of the adults we knew as a child and that of the adult we are among other adults and other children... There is no question as to how the childhood event acts only with a delay. It *is* this delay, but this delay itself is the pure form of time in which before and after coexist... What is originary... is not one series in relation to the other but the difference between series. (162–3)

Deleuze's theory is that what makes an adult react to the past is not the slow causality of the past, and not the retroactive interpretation of the present, but the overlay of past and present narratives. The difference between them—not the resemblance—defines both, and distributes them in time. Delayed reaction presupposes the pure past as the coexistence of distant events as well as the future as the desire within each event to communicate across time.

Deleuze here offers a subtle alternative to Freud's conception of 'delayed reaction', *Nachträglichkeit*, in *Project for a Scientific Psychology* (1895).[15] Freud's case study concerns a young woman who has a phobia against going into a shop by herself. Under analysis, she remembers that when she was twelve, two shopkeepers laughed at her clothing. After more analysis, she remembers that when she was eight, a shopkeeper grinned and tried to touch her through her clothing. Why can she not go shopping alone now? Considered as quantitatively separate events, present at distant moments in succession, the childhood and adult events would have to interact on the basis of qualitative similarities, for example, in that they both involve a girl and older men, clothing and grins. On the basis of those similarities, the girl would expect another similarity, another assault. But Freud does not think the similarities are strong enough to explain the intensity of her fear. Indeed, if the patient consciously generalized that 'shopkeepers are molesters', that would be a reasonable motive for avoiding shops. But the point is that she does not know why she cannot go into shops. Why is the reaction delayed, masked, and magnified? Freud's explanation is that it is only at the age of puberty that children realize what sexuality is, and at that age, they retroactively interpret their previous experiences in sexual terms. During the first experience, the girl was pre-puberty. At that moment, the event was painful, but not traumatic. At puberty, she realizes that, years earlier, the man was assaulting her. Worse, now that she has sexual feelings, she wonders if she had been complicit. The remembered event now becomes unbearable, so the memory is repressed. It becomes important and at the same time inexpressible; the 'simultaneous contradictory action'[16] comes out symptomatically as phobia. As Lacan says, *nachträglich* delay is a re-memory (*remémoration*) that creates a new reaction to the original experience.[17] The early experience cannot have its nature, or its effects, when it happens. It is 'premature'. It achieves meaning too late, because it occurs too early (Freud, 1966, 257).

For Kant, returning to the same place and finding it the same, defines simultaneity; for Freud, returning to the same event and finding it different, defines delay. This is delayed reaction by retroactive interpretation: an event that cannot be processed in the normal flow of anticipations and fulfilments, whose effects are excessively slow to manifest, and which later appear as a sudden onslaught of the past, as if after a delay the past speeds right up to the present without delay. Freud calls it *proton pseudos*, a false origin.

As it happens, Freud later changed his mind about key features of this account, possibly for bad reasons, but we do not have to worry about that. Delayed reaction clearly plays a role in many psychic phenomena, and the theory of retroactive interpretation seems plausible.

However, Deleuze's theory of time claims to render retroactive interpretation superfluous, construing delayed reaction instead as coexistence through difference. Deleuze and Freud agree that overlaying the difference between child and adult viewpoints, between innocence and desire, is what makes the early event traumatic (innocence violated by desire) and the later one phobic (desire blocked by regression to innocence). But for Deleuze, what dramatizes the memory is not that one now interprets the event through adult eyes instead of a child's, but that one now relives the event at both stages of life at once. Merely retroactively learning what a childhood event meant, or retroactively attributing adult motives to the child, would not make a person obsessively try to avoid it now, unless in addition one felt oneself a child again *while* going about one's present business. Conversely, one would not feel complicit in the childhood event, and try to compensate for it now, unless one felt that qua child one was always already forming and responsible for adult sensibility. Phenomena are synchronic in the sense that the same subjectivity struggles with them all at once, yet diachronic in that subjectivity is serially self-divided. To be sure, such a phenomenon could still be called retrospective. But for Deleuze, it might better be called reciprocal causality with reversible subjectivity. Delayed reaction is better called simultaneity at a distance.

Deleuze also argues that Freud's theory of delay erroneously depends on a 'solipsistic unconscious', rather than 'intersubjective unconscious' (162), namely the communication between the child one is and the adult one is. We might define schizoanalysis as a theory of internal intersubjectivity through delayed temporal effects across intersecting series of thoughts, memories, and affects. It is not hyperbole to say that when faced with an experience today, I run through one series of reactions that calls into play my eight-year-old self, and another my fifty-year-old self. Whenever a person walks into a store, a child is walking into the store and an adult is walking into the store at the same time. The primitive accumulation of childhood refers less to pre-historical events in chronology, than to a diagram of possibilities. The many subjectivity diagrams at my disposal constitute the coexistence of different moments of my personal history.

In short, the childhood series—looking up to adults, uncertainty about what things mean, powerlessness, and the struggle for self-identity—becomes entangled with an adult series—self-recognition, generalization, socialization, and sexual politics. The similarities provide only a minimal assemblage: girl, entrance, storekeeper, clothing, touch, grin. But once the two series are overlaid, then the childhood question, 'what do adults want from me?' is superimposed onto the adult question, 'how do I manage the desire of the other?' The two different perspectives together create figurative resemblances: 'shamefully, the storekeeper I encountered when I was a child was a man like other men'. Or, 'everything for sale is repellent'. The resemblance is a running effect of two different interpretations targeting the same object = x.

On Deleuze's model of delayed reaction, then, it is not that there exist two events separated in time, the potentials in the earlier event getting resolved in the later, or the later reacting to the earlier. Instead, there is one event occurring at two distant moments in time at once. In happier cases, the overlay need not be traumatic but could be the poetics of time regained, or intersubjective history.

Any given event might have a crossover point with an event stored in the past, whose gears it engages for a while. The force behind cross-temporal assemblage might or might not be voluntary. The connector could be economic, biological, technological, political, linguistic, or philosophical; a gift, a germ, a spur, a wave of immigration, a pun, or a concept (for example, the concept of delay) could set in motion a whole history of delayed reactions. The point is that delay is a structure of time before it is a matter of psychic, economic, or some other specific kind of development. It is not psychic repression that causes experiences to be delayed; it is the delay-structure of time that shapes the psyche into experiences like repression. Delay is any immediate coexistence of events that come before and after one another in time, operating as a single phenomenon. It might seem preferable to have a less general definition of delay, to explain each particular delay by a specific mechanism that prevents an event from occurring on schedule. And indeed, Deleuze intends to keep the assemblage of the event and its delayed reaction concrete. His works with Guattari are full of concrete models of delay, especially of the way capitalism uses regions of delayed development as lawless zones capable of sparking accelerated development. Yet sometimes, concrete mechanisms of delay, from retro-perspectives of puberty to apologetics of capitalism, are too specific. Concreteness can create the illusion that the delaying phenomenon, the transition to its successor, and the successor itself are distinguishable temporal phenomena. The more abstract combinatory art deploys delay as a universal structure of temporality. The happy product is that delayed coexistence allows us to build consensus with our selves and even suggests a kind of democracy without compromise.

Chapter 8

Simultaneity and Delay in Politics

Theme of the chapter: Decisions concern possible futures. Human history gradually works out synchronous events but staggers its continuity across distant points. The politics of simultaneity and the politics of delay are complementary.

For Deleuze and Guattari, state history (which they often refer to simply as 'history') describes an ordered succession of authoritative regimes, whereas revolutionary history (which they generally call 'becoming', but sometimes also call 'history') assembles contemporaneous series freely from across the historical field. But contemporaneous becomings cannot all occur in a single instant. History is diachrony within synchrony; or to say it in reverse, succession is intervallic simultaneity.

In the 'Apparatus of Capture', Plateau,[1] Deleuze, and Guattari discuss the historical encounters that allow capitalism to capture its past and at the same time allow something post-capitalist to escape (565). Capitalism appropriates the delayed meaning of its predecessors, but by uprooting labour from tradition, it inaugurates a delayed reaction against its own control. Deleuze and Guattari footnote Balibar. Balibar construes delay (*décalage*) as a consequence of complex synchrony, whereas their concept of delay (*retard*) evaluates synchrony as a variation on delay. Balibar discusses concretely the delayed emergence of, and then the delays within, capitalism. Deleuze and Guattari construe delay more abstractly. In their analysis, delays in economic transformation, and accelerations in the global free market, express the universal delay inherent in becoming. Deleuze and Guattari define all movement in terms of 'the infinite slowness of the wait … and the infinite speed of the result' (343, 'Becoming-Intense'). They cite speed and delay (*tardivité*) as diagrammatic traits of abstract machines (176–7, 'Regimes'). They say that 'the renunciation of external pleasure, or its delay (*retardement*), its infinite distance, testifies to an achieved state where desire no longer lacks anything but fills itself and constructs its own field of immanence' (193, 'Bodies Without Organs'). In short, '*retard*' and its cognates describe historical and other factual delays as functions of the delays of abstract desiring machines. Concrete details of history are already in some ways abstract, but it takes work to become abstract enough to engage history.

Balibar's philosophy of temporal politics compromises between the pure time of Deleuze's politics, the differential time of Derrida's, and the decisional time of Badiou's. Perhaps because he is more moderate (less ontologically pure), Balibar offers a great deal of material for a philosophy of historical delay.

(a) Delay as Diachrony to Come: Balibar, Deleuze and Derrida

Balibar

Balibar interprets diachrony in terms of structural variations following a synchronous situation. Occasionally, there are irreducibly diachronous transitions from one synchronous situation to another, but for the most part, what we call diachrony is the delayed operation of a synchronous structure. From an Althusserian starting point,[2] drawing occasionally from critical theory,[3] Balibar speculates on how serial contemporaneities pass through times of delay. His treatment of this problem over the course of many texts maps a history of delay theory.

Reading Capital[4]: diachrony and *décalage*

Classical historiography seeks 'the right break' between one historical period and the next but fails to see that breaks are 'structured' by economic modes of production (206). A 'structure' is a complex system of combinations and divisions (216), a 'relation of relations'. Structural systems explain most of what used to be called 'transitive causality' (224). What look like successive events are often just variations on a constant structure. Similarly, classical historians rely on a 'mechanistic thought of temporality' (236–7). For Balibar, as long as relatively stable structures operate, temporal changes play less of a role than those historians suppose, and when structures are in transition, there is a greater role for temporal change—and more varied types of temporality—than they suppose.

Balibar's Marxism is anti-evolutionary. 'The possibility of finding the periods of modes of production according to a principle of *variation* of combinations, alone deserves to hold our attention' (225). Differences from one state of society to another are largely just 'non-correspondences', structural rather than genetic. And when a significantly new phenomenon does arise, it is not just the phenomenon that is new but also its structure. The transition, for example, from animal-driven ploughs to steam-driven ploughs cannot be explained as a simple 'lineage', since the new plough has a wholly different function than the animal, being organized by a new system of labour and social relations. What looks like a small-scale succession (a new grain) could indicate a large-scale structural change (international trade patterns); what looks like a major advance (a national war of independence) could be structurally insignificant. It is better to speak of transitions as 'displacements' rather than in terms of 'continuity' (243) or 'development' (247). Strictly speaking, though, Balibar's Marxism is not exactly 'structuralist', since structuralism posits 'formal combinatory', whereas Marx's structures are related to modes of production (226).

Modes of production are difficult to define, since they consist less of specific technologies than of 'rhythms' and 'patterns' (235). A mode of production is a synchronic system, but it brings about 'chronological *dislocation* (*décalage*, or time-lag)' (236). Advances made in one industry are over time spread across different industries, labour pools, and legal systems.[5] A system like capitalism might employ its favoured modes of production 'simultaneously' (237), but temporal dislocations are part of how it extends that simultaneity. Delayed extension is essential to a structure's synchronic pervasiveness.

Just as there are time lags within synchronic structures, there are structural properties of diachronic transitions between one structure and the next. Not every succession is historical in the same sense, since each relates to its own structures. There is no 'history in general'. Even a phrase like 'the history of labour' wrongly assumes a continuous 'referent' or 'subject' (249), as if peasants and bureaucrats were labourers in the same sense. For this reason, historical analysis is inherently problematic, requiring decisions about which phenomena should count as the successors of which. And it is rarely obvious which structure is the key to a phenomenon: urban expansion may be a consequence of market structures or of governmental or policing structures. The 'different forms of historical individuality' (252) constitute 'not a multiplicity of centres but the radical absence of a centre' (253). Even though history is a series of synchronies, the designation of a given element in a synchronous state is deferred until we see which element in the following synchronous state corresponds to it. In short, the meaning of succession depends on the structure, but the meaning of the structure is underdetermined.

The only means of determining what succeeds what hangs on the 'repetition' of processes across time periods (259). If a way of doing something is repeated across successive contexts, that is evidence of a structure at work. Processes in the 'accumulation of capital', for example, are different across successive contexts on almost every parameter, for example, across different types of labour, currency, property, and banking. Yet in spite of being problematic, variation can help us identify a phenomenon across differences. For in general, reproducing a process in a new context forces old and new phenomena into commensurability (as old and new currencies can be added together and spent on a single commodity). Indeed, the paradigm phenomenon that is individuated only after it has been reproduced in new forms is capital. Capital is produced not merely by exploiting surplus-value but by self-perpetuating (263). A given act of exploitation is not really capitalist until it is repeated on a large scale. Reproduction synchronizes what would otherwise be serial diversity. Similarly, two events might happen at the same time, but if they have no structure in common, they may constitute only 'fictive contemporaneity' (264). In general, reproduction guarantees the 'transformation of succession into synchrony, into "simultaneity" (*Gleichzeitigkeit*)' (264). To make a long story short, it takes a lot of repetitions to bring a synchronous structure into play, but once the repetitions have occurred, it is possible to say that even the earliest iterations were part of that structure.

The problem of history is how a synchronous situation has a genuine history: 'how each of these times … *becomes* a historical time'. We said that two phenomena fall into synthesis only once individuated and combined synchronically. Only an eventual simultaneity can explain a succession capable of having lead up to it. Hence, we cannot explain in linear chronology how labour and wealth got combined in capitalism, since it is only after capitalism has been reproducing itself for a while that labour and wealth can even be defined. And once capitalism does exist as a synchronous system, the relevant forms of labour and wealth will already be operating together. It is misguided to debate which of labour and wealth is 'the preceding and the succeeding' (271). Once the synchronic structure exists, its elements, and their respective histories, already coexist (272).

The twist is that only once a result, like capitalism, exists all at once, is its prehistory, the accumulation of capital (a controversial subject in economic history, interpreted variously in terms of the rise of bourgeois productivity, the re-deployment of feudal or colonial wealth, new banking practices, etc.), genuinely successive. The successive accumulation prior to capitalism is an effect within the simultaneity of capitalism. Balibar does not construe succession purely by retroactive interpretation, which he regards as idealist. Nevertheless, transitions depend on synchronous structures. Balibar wants to affirm real historical transitions, even 'ruptures' (268), while denying univocal lines of succession leading forwards to periods of social history. He wants simultaneity to explain, but not abolish, succession. In my terms, he contrasts simultaneity not with succession but with delay.

Furthermore, Balibar wants to explain structural transitions without a general theory of transition. Different transitions, and their before-after relations, have different structures. To that end, he describes two cases, and two definitions, of transition: (1) the 'pre-history' of capitalism in 'primitive accumulation'; and (2) the 'tendency' towards the dissolution of capitalism. In analyzing the former, Balibar critiques retroactivity; in analyzing the latter, he affirms temporal pluralism.

Transition 1: prehistory vs. retroaction

The primitive accumulation of capital begins only once the threshold towards capitalism has already been crossed. It is a myth to say that capitalism gradually developed out of a pre-capitalist right to property, for example, and that only later did owners exploit labour-power. Balibar calls this 'retrospective projection' (276), culturally subjective 'memory' in service of capitalism's self-justification, rather than genuine 'history' (278). There are no genuine origins of capitalism prior to capitalism itself; the origins of capitalism fall within capitalism's own early stage. Its 'pre-history' is part of its history, not prior to it (278). The twist in Balibar's account is that only if an early stage is taken to occur *within* a period can that stage be held distinct as prehistory from the mature stage within that period. Any claim that an early stage occurred before the period, is likely a retroactive apology. Retroactive interpretation is admittedly appealing, since it sees correctly that earlier stages of capitalism occurred while its elements were still relatively independent, for example, when finance capital, labour exploitation, and so on, made up a still undecided 'diversity of historical pathways' (282). But what is wrong about retroactivity is that it ignores the 'find' (*trouvaille*) whereby independent processes were at a given moment forced into a joint result (282).

At first, it may seem that in rejecting retrospective interpretation, Balibar affirms succession: an event that really took place before capitalism. But Balibar's subtle point is that capital accumulation took place *before* capitalism, but *also during* capitalism, and that this type of paradoxical sequence, neither a-historical rupture nor normal occurrence within a period, is what we call 'transition'. In my terms, a transition is neither pure succession nor pure simultaneity, but a delay-event. Balibar would not say, as Deleuze does, that the same event occurs both during childhood and during adulthood, but he might say that the adult's past occurs at a moment when one's adulthood is still in its childhood state.

In sum, it is deceptive to use retroactivity to claim either that pre-capitalist build-up was an innocent process prior to it or that all societies have included capitalism to various degrees. In contrast, it is right to identify prehistory as the moment at which non-capitalist processes are suddenly brought into a common result and a common history. These two themes—the plurality of genealogical pathways and the structural synchrony of their effects—will generate Balibar's theory of time.

Transition 2: double references and décalages

The eventual dissolution of capitalism begins as a 'tendency' (283). Every structure is in 'movement towards its historical future', simply due to 'its existence in time' (285). To be 'in' time does not mean falling under the 'chronology of a succession' (285), but rather that a process plays out in phases. The law of process determines its dissolution 'in the long run'. The theory of the 'long run' implies 'retarding causes' (286), namely causes that specify, yet delay, indeed specify as delay, the rhythm of the process's 'counter-effects' (287). 'Synchrony (Marx calls it a "simultaneity")' is the 'double effect' (*doppelseitige Wirkung*, 289, or *zweischlägtig*, 290) of simultaneity and delay.

Balibar takes up Althusser's call for a new theory of 'historical temporality' (293). Too many philosophies of history take for granted an abstract philosophy of time. They assume time is 'continuous, linear, and unique', 'order, unfolding, and duration', pivoting on the 'contemporaneous present' (294). They assume that events occurring at the same moment in time 'belong to the same history' (294). The most they can say regarding temporal diversity is to contrast short-term events with long-term stabilities (wrongly called 'structures', Balibar adds, 294). The classical theory of historical 'periods' assumes this model, thereby excluding 'true diversity' (294). Of course, even naive historians notice different tempos within a mode of production (for example, that a day's labour may be longer or shorter than the time needed to produce a commodity). But their descriptions merely 'insert different times one within another' (295), for the sake of forecasting. Theoretically, their reduction of rhythms to homogeneous time-segments is neither diachronic (since it excludes novelty) nor synchronic (since it measures empirical successions rather than systems).

In truth, even within the same synchronic structure there can exist essentially different rhythms of succession: the generation of profit includes the 'time of social labour' with its rhythms and timescales of productivity; the turnover of capital includes rhythms of circulation; crises of overproduction accelerate with transportation time (298–9). Multiple timescales within a single period make it 'dynamic'. They also permit 'backward' and 'forward' development, as when two processes occur successively in one nation and simultaneously in another (300).

If differential temporalities help to describe synchronic situations, they are crucial for understanding diachronic transitions. The twist is that the temporal character of diachrony is thus like an especially intense case of the differential temporal character of synchrony. With the right kind of '*differential analysis*', whenever we see inconsistent functions coexisting (synchronically), we say that a (diachronic) transition is taking place (307). A period of transition is a 'double reference to the structure of two modes of production' (307). This is the essence of the 'theory of *décalages*', or time lags. A period of transition is '*the co-existence of two modes of production in a single "simultaneity", and the dominance of one over the other*' (307). Methodologically, 'problems of diachrony must be thought within the problematic of theoretical synchrony' (307).

This theory of coexisting yet differential time-lines is helpful in accounting for historical (and indeed all) phenomena. But given differential time-lines and lags, why does Balibar even need a division between synchronic structures within a historical period and diachronic transitions between them? Balibar thinks that synchronicity is a necessary condition for the coexistence of different time-lines, because only a system can coordinate rhythms of labour, production, and circulation. But the transition events, like accumulation-encounters and tendency-generators, seem by themselves capable of juxtaposing differential temporal determinators, whether structural synchrony is posited or not. For that matter, those temporally differential processes seem capable of accounting for the very difference, if there needs to be one, between structural and transitional processes.

If the division between synchronic and diachronic phenomena is legitimate, it supports a distinction between phenomena appropriate for their own period and those whose usefulness is delayed. But if that division can be tampered with, it might free up delays within simultaneity, and free up simultaneity during and across diachronic transitions. The paradox for Balibar's solution is that once periods are explained by means of diachrony and synchrony, and diachrony and synchrony are explained by means of simultaneity and delay, and simultaneity is explained by differential temporality, namely systematic delay, then delay turns out to be the only one of these categories that is finally necessary. And yet, once delay is given its due, the picture of history looks different. In later texts, Balibar extends the role of delay into more uncertain futures, into periods before transitions take off, and sub-stages within synchronous periods. He also introduces subjective and epistemological delays, alongside socio-temporal delays. All of this approaches a universal theory of delay.

The Philosophy of Marx[6]: dialectic and deferral

Why has the communist revolution been delayed? Balibar discusses two sorts of delay: the delayed telos, uncertain, contingent, and deferred; and the delay that assembles variations between the end of one system and the presence of the next.

Balibar critiques those capitalist economists who read pre-capitalist societies historically but read capitalism ahistorically (81). He equally criticizes 'postmodernists', who ridicule progress yet assume their own views are progressive (88). Balibar wants to defend historical improvement without relying on teleology.

Again, history does not follow one 'model' (96). It is true that dialectical materialist causality, the idea that 'in the last instance' an economic base determines a superstructure, is abstractly teleological. We might even say that ideology is a delayed reaction to material conditions. But what base structure produces is precisely class struggle: multiple simultaneous possibilities, a situation with no single telos. Struggle is not a 'general form' of progress, but 'differential' 'accelerations' of forces (101). The paradigm is the double status of proletarians: as workers, they build capitalism, but as revolutionary agents, they dissolve it. There are historical contradictions largely because delayed results are simultaneously built into the processes that precede them. The doubling of a single phenomenon, the multiple dynamics coexisting within a single moment tending in different directions, makes the history of struggle inevitable, whether a certain plan succeeds or not. This differential makes it impossible for capitalism to reach ideal stasis, and so implies its dissolution.

The 'endlessly deferred impossibility' of capitalism not only leaves room for a proletarian revolution, but also generates historical temporality (101). 'Historicity' results from the contradiction between production and its impossible fulfilment. Historicity is the opposite of narrative finality (102); it is the difference between a force of production and itself; it is immanent delay. Yet while Balibar suggests that history is universal delay, the rest of his text confines delay to local issues.

Balibar focuses on two problems that have faced Marxism during the long delay between its first enunciation and the allegedly immanent revolution (103). First, why does Marx require the dictatorship of the proletariat as a necessary stage in the transition to communism? Second, why does Marx require that industrialization occur before communism? Is there not evidence that these transitional stages do not facilitate, but rather extend the delay of, revolution?

At the empirical level, Marx can answer that the proletariat needs time to become self-conscious; it needs a transition within a transition, a will to diachrony while capitalism is still synchronous. No doubt this is plausible. But what is the quality of transition? In revolutionary history, it is 'as if time were to open up or stretch out to make the way, "between" the present and the future, for a practical application of the "classless society" in the material conditions of the old one' (106). This is closer to a Derridean motif, where a question opens revolutionary delay, than a Deleuzian motif, where temporal assemblage is itself revolutionary. What is important for Balibar is Marx's insistence that intra-transitional phases, even if they increase suffering, are necessary just because between immanence and progress there is 'provisionality': 'historical time's "non-contemporaneity with itself"' (106). Balibar's final statement that some societies might need transitions that others do not, is empirically plausible but theoretically mild. The structures of delay should lead beyond apologies for slow progress and move towards a theory of revolutionary delay as such.

Race, Nation, Class[7]: genesis and hesitation

As Balibar adds genetic analyses of history to structural ones, inserting delay at more points, softening his earlier anti-evolutionism, we find a reduced role for simultaneity. Delay no longer exclusively means the staggered double reference of a synchronous event, but also means the undecidable result of genetic succession.

Balibar's concern in *Race, Nation, Class* is the history of nationalism. The narrative of the 'nation' misleads in assuming a continuous project of nation-building through many generations (86). In reality, there are multiple origins of a nation, with interactions across language, governance, currency, church, law, and police at different dates. There is no single founding phenomenon, but a 'threshold of irreversibility' at which categories of nation become a 'system' (88).

Though Balibar says we cannot 'deduce' the nation from one phenomenon alone, he emphasizes the proletarianization of labour. He says (as do

Deleuze and Guattari) that the division of labour divides 'core' and 'periphery', and so divides nations from colonies. For Balibar, different historical pathways, in their simultaneous potential, are not predetermined: tribal as well as intercultural societies compete with early nations, and early bourgeois classes 'hesitate over' whether the nation state is the best way to rise to power, control armies, and exploit rural populations (90). This hesitation among options explains the 'time-lag' in the emergence of the nation in the early days of capitalism (90). During the delay, there exists a 'simultaneous genesis of nationalism and cosmopolitanism' (90). Delay is identical to simultaneous possibilities competing for hegemony.

Balibar does not classify delay-forms, but we can find many in his text. (a) When pressures favour non-national ethnicity even after nations become prominent (90–1), we could say that delay-as-anachronism follows from delay-as-hesitation. (b) When Balibar asks, 'for whom today is it too late' to become a nation? (91), wondering whether de-colonized societies can become nations if they have no single language, no protected markets, and no backward peripheries to exploit, we could say that after too much delay, no further such delays are possible. (c) When Balibar emphasizes that nations are incompletable, that even old nations suffer 'delayed nationalization' in the form of 'an endless task' (92), for example, the endless task of excluding immigrants, we could say that delay-as-instability and delay-as-incompletability follow from delay-as-contestation.

It is now too late to believe that nationalism can succeed. Having been delayed, it is too late for the nation to arrive at some later time. Of course, we have to admit the relative stability of capitalist nations, and the success of its narrative, even if its threshold is never met (92–3). But the narrative is forced to dip into diverse resources, from military glory, to census statistics, to collective interpellation. The nation does not just 'territorialize', Balibar says, but de-stratifies pre-existing relations in order to reconstruct them into an imaginary signifier.

Balibar ends with the remark that we cannot predict whether European nations in the future will stabilize, reform, or dissolve. No doubt this is correct. Nevertheless, multiplicity historiography and hesitation subjectivity should lead to a broader logic of delay.

Masses, Classes, Ideas[8]: delay during the delay

In *Masses, Classes, Ideas*, the locus of delay is the time remaining (see Agamben) within the age of capitalism before the revolution. Marx criticizes bourgeois historians for failing to imagine history after capitalism, but

Balibar criticizes Marxists for failing to see history *during* late capitalism, namely, during the time in which capitalism's dissolution into communism is delayed. Balibar begins with why Marx only rarely names the 'proletariat' in *Capital*. The short answer is that while the bourgeoisie has a 'historical character' throughout capitalism, the proletariat only takes up this role towards the end of capitalism. Before that, the proletariat plays an amorphous role as the 'non-subject of history' (142–3). Even when the proletariat does emerge, its 'class' structure (self-conscious alienation) must still become a 'mass' movement (as the agent of history) (144–5). Attempts to explain the agency of the proletariat during late capitalism (via ideology or via the Party) too often underplay its transformations (e.g. the stages of unionism) (160). In short, some Marxists posit the historicism and transience of capitalism, but do not trace enough history within it. They affirm 'historicism without history' (148, 156–61).

Balibar says that a theory of the masses as '"motor forces" of history' must go 'much further than the simple idea of "retroaction" (*Rückwirkung*)' (156), further than 'Althusser's "action at a distance", or "absent causality"' (157). It is not enough to promise that the proletariat will have been guiding history before it was unified. We need to know what happens between events, in unfulfilled programs. The best historical materialism should be incomplete (174).

Balibar invokes historicity *with* history, a description of historical events during the period of delay. Deleuze is more interested in undecidability than incompletability, in preserving inclusive disjunction rather than waiting to see which disjunct prevails. Nevertheless, Balibar shares with Deleuze the view that history is the in-between of historicity and that neither retroactive interpretations nor distant former presents with slow fuses; in short, neither the afters nor the befores explain the delayed motors of history.

Balibar has developed three broad roles for delay. First, multiple time-forms operate in the same historical period. Second, delay governs diachronic transitions between periods. Third, delay covers the incompletability of synchronous structures and the unachievability of their supposed successors. We could say that the first delay shapes multiplicity, the second chronology, and the third finitude, which all together shape historical *différance*.

Balibar's philosophy is intended to provoke action, solidarity with the excluded, and up-to-date reflection on the global situation. But we need to know how intervention in permanent delay might work, how we might find momentary openings in a time-line that does not have moments as such, and whether there are exceptions to the delay of all projects. We need to know how we might use the very pervasiveness of delay to change something—to perform not delayed action but active delay.

Historiography (the theory of transitions) *with* history (in its in-betweens) is not limited to political history in the narrow sense. We find an illustration of political action-delay in Deleuze's work on cinema.

Deleuze

Whereas Balibar emphasizes elements within synchrony that spark delayed reaction, producing diachrony before, during, and after transition, Deleuze (following Bergson) finds synchrony itself to be a kind of delay, so that temporally distant events occur as a singular event with multiple coexisting historical pathways.

Cinema 2[9]: too late

Films that represent historical events (not counting direct cinema broadcast in real time) do so only after a time lag. Can this delay contribute to its political efficacy? On the surface, the question leans towards apolitical politics. How can one contribute directly to an event if the event has to be put into art, or concepts, first, after the fact? Does delayed history imply that we cannot genuinely act on events as they occur, and that our motivations and strategies come too late? Deleuze is sometimes criticized for his Stoic *amor fati*, as if he is content with whatever happens in the body politic. His reply is that the late arrival of events does not make us passive. If arriving too late is 'dynamic', the too-late structure is just what allows us to act. In simple terms, delay allows us to contemplate before acting. And the simultaneous coexistence of distant events allows us to have an effect on what we saw earlier.

In the last pages of the 'Crystals of Time' chapter of *Cinema 2*, Deleuze says that in Visconti's films, and later he adds, in Syberberg's, '*something* arrives too late' (*trop tard*). It's the SS; it's Hitler. It may not be possible to experience a historical event exactly as it happens, but cinema represents how events are represented. 'History growls at the door'. We do not exactly experience a historical event in cinema, yet history is substantially present in recorded images, not just as 'scenery'. History confronts us, just too late. If it could have been 'caught in time' (*pris à temps*), we might have averted it, but 'it is History, and Nature, itself, the structure of the crystal, that makes it impossible for something to arrive in time' (126). It is not that Visconti is nostalgic or backward-looking, it is that looking backwards is the only way history can be apprehended. History is an 'autonomous factor', and being too late is its 'rhythm'. Delay is 'a dimension of time', the universal '*Nevermore*'.

And yet, 'history arrives too late *dynamically*' (127). The paradox of the event that appears not in its time but in representation is that it is both harder and easier to avoid. It (fascism, capitalism, or childhood) arrives too late, so there is no time when we can prevent it, so it remains forever. But on the other hand, since it is too late, it never was where we look for it, there is no one point that makes it definitively or empirically factual. So we do not have to keep finding it, and it no longer has to be there in the guise of inevitability.

Because of its dynamic, the too-late has three quasi-exceptions. First, art thrives on the too-late arrival of our ability to see it. An event occurs too early for the human, phenomenological eye to see, but as Vertov would say, the camera eye can see it, and by extension, we might say, the History-Eye can see it. Second, the too-late succeeds at preserving time itself, 'time regained' (128). As Balibar would say, we never control the event when it occurs, but we regain it when we reproduce it by selecting a subset of its possible consequences.

Third, Deleuze says surprisingly that when cinema sees the event too-late, it 'defeats Hitler' (352–3). What is Deleuze saying? It is too late to defeat Hitler. But what is Hitler? Strangely, Deleuze assesses Hitler in a way similar to that most famous too-little-too-late text in philosophy: Heidegger's 1966 *Der Spiegel* interview,[10] characterizing the problem with Nazism as its inauthentic communications technology, its reduction of discourse to information. Surely that says too little about Hitler. But Deleuze says it too: Hitler succeeded by convincing his people that what he said was what he said it was, that what he said was, that what was was what he said, in short, by creating the cinematic illusion that expression is simultaneous with content, by synchronizing sound and picture. Defeating simultaneism requires a different kind of cinema, one that creates delay between speech and the visible, making each too late. Perhaps this counterattack of the too-late should have been caught at the time. Some Godard should have made *Medium Cool* at Nuremberg. In fact, Ernst Bloch argued in 1935[11] that Hitler was harnessing the power of 'non-contemporaneous' social forces, and that his opponents had to defeat him with their own non-contemporaneities. Authorities cannot act on us if they cannot reach our nows, and if we can make a lifetime in the interstices before they reach us. It may be an accident that the too late was discovered too late, but it is not too late to discover it now. It is not too late to make history, but to do it is to unweave the one-at-a-time event-units that claim a thousand years. To make history now is to make an act that *is* the delay, an act in which the now is not merely now but simultaneously stretches along the staggered time that it is not, to act on trauma and proletarianization in the dimension of the too late, though of course, hopefully, in time.

Why is there always one extra transition before a revolution can occur? Why is history one long hesitation? If the answer is that even synchronous situations are filled with non-contemporaneous dynamics, then delays are not so much obstacles to, as forms of, decisive events. Overlapping at a distance, or simultaneity at a delay, will be a strategy for defeating dictators, a therapy for childhood traumas, an alternative to capitalism's rewriting of history. This only works if synchrony doubles itself with internal time-differentials, and if diachrony doubles back the externality of succession. Balibar worries in a recent text[12] that unless the 'rupture of time' is 'determined in the last instance' by material history, it may lead to something more 'demonic' than revolutionary. Deleuze may be less worried about demons than Balibar, but the ongoing challenge for the philosophy of history is to show how concrete assemblages of abstract temporalities can do the work that revolutionary historiography used to do with succession.

The most extensive politics of delay belongs to Derrida.

Derrida: politics of delay

Derrida's politics are governed by delay. Badiou's are governed by simultaneity. But since delay and simultaneity govern each other, Derrida and Badiou represent the two sides of the dialectic of temporal government.

In *Rogues*,[13] Derrida appeals to his earlier deconstructions of 'simul', to argue (not especially controversially) that democracy is not a settled constitution of social relations, electoral processes, institutions, or popular decisions; it is a way of opening discussion. Derrida defines 'democracy to come' partly through his earlier themes—equality without calculation, singularity (national and individual) without auto-immunity, and justice as responsiveness—with new elements involving rogues, eccentrics, roués, turnings, and divergences.

When Derrida defines *à-venir*, 'to come', he spends a lot of time preventing misleading interpretations. He says that democracy is aporetic, ongoing, and open to revision (except perhaps about justice, deconstructively construed). Like friendship and hospitality, there will never be a time when we can count up democracy's successes and failures, and judge whether it has worked. Derrida's formula for democratic discourse is: 'militant political critique without end' (126). This does not just mean that democracy 'will never be perfect' (since that concession treats perfection as a telos whose meaning we already know, and know to be beyond human achievement, 108). Democracy should be possible, not merely a hope, a wish, a prayer, or a utopia (112). Derrida says that democracy is an 'infinite argument' (29), effectively delaying decisions indefinitely. Attempts to limit discussion constrain democracy with the demand for 'sovereign decision' (38). Decisional finality 'tends towards simultaneity' (32). It is not only in certain cases that counting ballots is an endless process that only a court's fiat can cut off. It is not only that in certain cases, delaying an election at a time of public anger might save democracy for later, that is, save the electorate from making a decision that would destroy democracy (54). Certainly, delay can be a successful tactic. Lenin cites the Roman General Fabius Cunctator, 'The Delayer', for whom the Fabian Society was named, known for procrastinating tactics and avoiding decisive battles.[14] But Derrida's emphasis is less on tactics than onto-deconstruction: democracy to come 'gives the time that there is not' (153).

It is not as though every political order whose temporality is 'to come' is automatically good or democratic. *Rogues* begins with the post 9-11 states that become as rogue as the political organizations that they condemn for their roguery. Derrida calls this the worst of the to-come, because it 'comes from the future' (149) to disrupt what was good about the traditions of democracy. The good to-come is not particularly futural—both because it exists at the pleasure of what is, and because it is the only hope for what has been. But what can we say positively about the democracy to come that gives time that there would not otherwise be?

Derrida says that in the to-come, 'a certain annulment of time demarcates itself' (154). After citing commonsense examples of how democratic communication is not literally endless (that some decisions are made by deadline; that others expire at a pre-ordained date, 154), Derrida's discussion of the dialectic of time and its annulment takes an odd turn. He turns to Heidegger's excuse, during the *Der Spiegel* interview, that 'only a god can save us' (158). When the reporters give up asking Heidegger for an apology for his wartime past, and ask him instead what the politics of 'our time' ought to be, Heidegger replies that we should question the meaning of time itself. This might not be a bad return, both for Heidegger, who at that point has few options, and for Derrida (who like Plato is not adverse to quoting collaborators for words that republics can use better), who is not fooling when he says that politics is a form of temporality. Still, for a reader who wants to analyze the temporality of the to-come, there is not much actual questioning of time in this text.

To make headway, we have to invent some basic features of the phenomenology of decision, what I will call the temporal noema of a decision—not the subjective sequence of decision-making (urge, deliberation, decisive moment), but the temporal character of the objective situation that the decision makes a decision about. The tricky thing is that decisions are clearly about creating a new situation in the future, but they are also about the current situation to be changed. They are about determinate future states, but have to remain flexible enough to hold good as circumstances change, plans fail, or understanding improves. Of course, peoples' decisions are often made in concert or competition, often simultaneously, that is, before each knows the other's decision, as in the Prisoner's Dilemma, played over a series of simultaneous rounds (in 'delay supergames'[15]). The challenge of rationality is not just to predict in the face of uncertainty, but also to balance present and future values.

Decisions refer to what one intends to do in the future, but the specific moments in the future to which they refer are impossible to limit. They are saturated with virtual futures. Their future referents are in one sense given simultaneous with the original decision, and in another sense, delayed. The openness to different possible applications, based on how the world happens to turn out, suggests that the time-line in the decision's objective referent is set off from—parallel to, but not—the 'real' future. And this suggests that delay refers not just to a later moment of time, but to an autonomous temporal series that we could call delay-time, a virtual phase shift from actual time, yet which nevertheless has to join up again with actual time at those moments when the decision is at some point actually put into effect. This is the interesting twist on decisional time. If political decisions are open and delayed in this way, politics has an alternative reality for its real object, a reality that is not a utopian vision for the end of time, but works on the possible worlds that inhabit the alternate temporality coexisting in present intentionality.

A decision is not an idle promise to oneself, but a commitment. It sets a new sequence of acts in motion. But which acts, when? The moment of decision is paired with those later acts that, in retrospective fidelity, as Badiou would say, re-enact it in practice. The nation-founding decision that declares citizens equal is grouped in a set with all of those moments in the future of that nation when further decisions have to be made about equality in relation to illegal immigrants or enemy combatants. For Badiou, a decision is defined, evaluated, and owned retroactively. But I want to focus on the decision's forward reference.

If I decide to wake up at 6 am tomorrow, there is no puzzle about the temporal referent. If I decide in general to start waking up earlier, there is an open set of referents—various morning times on various days—but still fairly closely circumscribed. But if I decide to spend more time with my family, or to do something about global warming, there is no specific action, and no well-marked range of actions, that decides success or failure. The decision is clearly void if I make no attempt at all to carry it out. But which future events are intended by the moment of decision?

Consider what the decider figures the world will be like once, and while, the decision is carried out. On the one hand, decision aims to make the world different in the future; it aims at something different from the actual. On the other hand, the object of our decisions is the real world, even though deciding is not the same as predicting either one's actions or the world's future. Being decisive may even put the future more in doubt.

A decision is neither non-worldly nor worldly. It is about a quasi-actual world. And it is neither hopelessly delayed in temporal indeterminacy, nor synchronized with future events. Some propositions are neutral with respect to time, like mathematics; others do predict future events. But decisions are not like either. Their noemata are temporally variable, in-between imagined and actual time.

If I decide to do what I can about climate change, I do not expect my first recycled can to cool down the world. I expect the world to continue pretty much as it would have otherwise, with a miniscule difference that may or may not contribute to larger divergence down the road. I look forward to a temporal succession in the real world that shifts just slightly act-by-act. Once I come to my decision's second enactment, and its hundredth, there will have been uncountable unforeseeable divergences. And yet this differential is what the decision's future is about. The decision I make now is intended to affect an actual future, to treat actuality as a field of possibilities, and turn that possibility into a new actual further in the future, which it treats again as a possible, and so on; but it does all this now.

Decisions rely on possible world scenarios, but they need real material futures as intentional referents to diverge from and to work with. I cannot decide to do something if the referent is too variable. I (probably) cannot decide to fall in love with whoever is the next person to get off the bus. But the hook to actuality is not just a matter of repeating love in a new setting. Decision and love are projections of possibility through actuality back *into* possibility. It is not quite that I do not now know what I am deciding to do. It is more that in decision I live through a series of ready possibilities that at each beat will be exactly what the actual world is at that time; but at that same time, each beat will phase out of what the actuality will be into what it should be. A decision has to re-insert itself into a world that has been changing since its last engagement. To be sure, at each point, the decision makes something actual. But as Bergson says about memory, once decision is carried into action, decision ceases to be the primary phenomenon. Decision arrives one moment too late to have been actual and too early to be actual yet. The time-line of the decisional future is at each point out of phase with actual time by the length of just one instant.

Of course, there are also delay-structures in the subjective noeses of decision-making. Sartre's description of his decision to go on a diet ferrets out all the bad faith fits, starts, and downright postponements that are part of the decision from the start.[16] But noematic delays are just as immanent in decision, and more usable.

To analyze how an event to-come comes, we would need a lot more concrete virtual logic of delay interaction. Despite Derrida's excellent, sometimes temporological, analysis of concrete political situations, he is not focused on the machinic aspects of deferral assemblage.

At times, Derrida's democracy to-come that is always already our responsibility, which he sometimes calls messianism without a messiah, seems like a case of delay without waiting. Derrida criticizes Heidegger for undervaluing vulgar clock time in favour of profound ontological time, for suggesting that the actual succession of phenomena does not matter. But it is possible that Derrida's to-come likewise makes insufficient reference to an actual future, skipping over actual amounts of successive time. Marx, after all, thought that a revolution could actually take place, after certain concrete preparations, and that after that, things would actually be different, and new processes actually begin. Marx did not just aim at a new dawn, futures that never arrive, as Nietzsche and Heidegger (or worse, Hölderlin) do. If longings were all that could be intended by concrete delay, then Derrida would be right to say that we might as well take responsibility for the future now. But it is not out of the question that actual changes some years from now could change the politics to-come. Space travel, for example (though philosophers since Arendt tend to treat it as a diversion from political relations), might generate new geo-politics, new communications ethics, new models of distributive justice, even new interlocutors. If it were not for the loss of nerve, curiosity, generosity, and worlding of humanity in the late twentieth century, we might have been choosing this outer spatialized future as we speak, one of the many forms that the otherness of the to-come could take in real time.

The problem is that if no determinate social changes make a situation more rigorously democratic than it was, and if no quantity of time, filled with the right quality of events, ordered by the right sequences, and covering the right scope of potentials, shapes the delay of the to-come, then Derrida's politics of delay has already arrived without delay and there is nothing actually left that has not come yet. Politics rather needs to imagine what we have not yet begun. We need to know about the political machinery of democratic delay, and how it can be powered up starting now.

A few years ago, an MA philosophy student ran for municipal office in Guelph under a deconstructive ticket. He devised platforms based on a politics of deferral, including positions on low-cost student housing, zoning for 'big-box' Walmart retail stores, legal requirements for running for office, formats for campaign literature, the interpretation of paper ballots, and procedural rules for city councils.[17] One way or another, the undecidable to-come must be affirmed in advance in the time of politics as-is. Now is as good a time as any to count as to-come, but what is simultaneous with the now has to be genuinely not now. This makes Derrida's politics of delay not so different from Badiou's politics of simultaneity.

(b) Simultaneity and Decision: Badiou

For Badiou, political decisions are made all-at-once, without the delay of deliberation.[18] But if Derrida's delay arrives without delay, Badiou's simultaneity arrives non-simultaneously. The politics of simultaneity and the politics of delay work together just when simultaneity and delay undercut each other.

Badiou's philosophy of multiplicity begins retroactively—'the multiple is retroactively legible as anterior to the one' (*Being and Event*, 24).[19] Any thought or collection presupposes a decision about what to count together, so every thought is a result, and none is foundational or natural. Normal politics obscures this procedure, excreting cover-up mythologies to make it appear that the state collective is historically grounded and conceptually inevitable. In truth, political events occur when by decision we count elements that were previously invisible (e.g. count women as voters, illegal aliens as citizens, proletarians as comrades). To add more elements, we subtract rules for exclusion. An event in political history occurs when the absence of sets of subjects is noticed and corrected. History is the advent of abnormal multiples.

Temporality is indirectly implied when the decision to count elements is followed by a second decision to continue doing so on principle. This is the second retroactivity, which in a sense comes first (as Balibar and Derrida could agree). In Badiou's paradigm, it is only in the midst of the French revolution that its cause, namely the principle of counting peasants as citizens, takes shape. One could never have proved that peasants were capable of acting as citizens before they do so; one could never say that the cause of the revolution was the thought of citizenship (rather than resentment or hunger) until it is articulated *in medias res*; one could never say that the event is a revolution at all (rather than anarchy instigated by foreigners) until the events mature sufficiently to name themselves as a singular event. 'It is the revolution which has since retroactively given meaning—by being inscribed, via decision therein—to that historical situation that we call France' (203).

Badiou appeals to a distinction in set theory between belonging to a set and being included in a set. An element, or a subset of elements, that 'belongs' to a set is present in it. If a set is 'included' in a larger set, all the members that belong to (i.e. are present in) the smaller set also belong to (i.e. are re-presented in) the larger set. It is possible for a subset to belong to a larger set without being included in it, if the members that belong to the subset do not belong to the larger set. To use Badiou's example, a family may belong to a society, but not all of its members belong to the society (e.g. if some of its members are clandestine rebels, or illegal immigrants, or women excluded from voting). Normally, a set cannot be a member of itself (otherwise the Russell paradox destroys the consistency of set theory); what defines the set cannot be inside the set. But the mythology of state authority claims just this, namely that the state is the exclusive author of the reasons why it contains all, and no more than, the members that it does—in short, that what defines and grounds the state is inherent in it.

To summarize, (a) a set, for example, the set of members of a society, can be forced to include members that it did not include before; (b) knowing the pre-existing definition of the set, we cannot anticipate that it will later include these new members. The openness of (b) defines the 'pre-evental' situation; the novelty of (a) defines the 'event'. There can never be a proof that by definition the new members belong to the situation, there can only be a decision. Temporally, there is no continuous transition from the pre-evental situation to the event, there is only an 'interval' between the things in the situation and the decision that defines it (208). Walter Benjamin says that revolutions bring new calendars, that it is not an accident that in the July revolution, 'the clocks in towers were being fired on simultaneously and independently from several places in Paris'.[20] Like Plato's Great Year, simultaneity is a point when the calendar recommences. This lets us describe the relation between simultaneity and delay in Badiou.

As a decision, novelty arises all at once—not gradually as new ideas spread, but at one blow, when something previously regarded as absurd becomes real. Grounds for the decision become apparent simultaneously with the transformative event, and not before. However, this simultaneity is intervallic. The members who suddenly now belong to society do not in this case arrive as immigrants; they were already there. But their already-thereness is assigned *after* they were already there. Badiou does not say that events are spontaneous 'commencements', with no material conditions pre-inscribing them; he disparagingly calls that theory 'leftism', which is almost as bad as the conservative attitude that nothing new ever happens. To avoid both materialist conservatism and idealist decisionism, Badiou posits 'evental recurrence', 'evental between-two', intervention in which one event is 'assigned to an *other* event'. To this extent, he allows reluctantly that the theory of intervention is a 'theory of time' (209–10).

From the indemonstrability of the intervallic decision, a second interval follows. It will always require yet another assertion to verify and interpret the first—a new set of researches, a new organization of known elements. And since the motif of expanding set-membership (adding new voters to a society, new artworks to a museum, or new infinite numbers to mathematics) both repeats the act of subtracting restrictions, and also challenges the meaning of the expanded society, art, or ontology, it will always require still another affirmation in order to find the previous event well-founded. These two aspects of 'fidelity'—repeated affirmations, and verifications of interpretation—constitute the 'discipline of time'. Badiou calls politics the 'organized control of time' (211). Like Derrida, he says that the discernment of the truth of an event is always still 'to come' (397). And like Derrida, Badiou says that the elements that make up history (398), as well as the subject who knows it (401), are present at the time only in the sense that they 'will have been'.

Badiou is deliberate about the formal structure of the retroactive procedure for determining the truth of an event-claim. The difficult last chapters of *Being and Event* use the mathematics of 'forcing' as a model.[21] Paul Cohen developed this mathematics to solve the problem of the continuum hypothesis. After Cantor, we accept that there are infinite numbers greater than the infinite series of rational numbers. One large infinite is the power set of all combinations of the infinite set of rational numbers. Another is the set of irrational numbers. Another is the transfinite generated by Cantor's diagonal proof. The question is whether there is an ordered continuum of increasingly large infinite numbers. Is it always possible to determine whether one infinite number is larger than another infinite number? Is it possible to exhaustively list all the infinite numbers between two given infinite numbers? For a long time, it was not clear whether the hypothesis that infinite numbers can be plotted on a continuum is implied by, consistent with, or inconsistent with, the (Zermelo–Fraenkel) axioms of set theory. Cohen's method (beyond the scope of this author to analyze) is to subtract everything inessential from an original set (in this case, the Z–F axioms), then to add additional members to the set (in this case, the continuum hypothesis), and to assess whether the new set, with the new members forced into a set with the old, is consistent. Cohen finds that the set of Z–F axioms is consistent whether the continuum hypothesis is added or not; therefore, the question whether the continuum hypothesis is true or false under set theory is undecidable. Badiou's application of this method is intended to decide whether a historical event is or is not consistent with a certain way of being faithful to it retroactively, to systematically enquire whether there is a continuum of follow-ups to a historical event, to reduce a historical event to its minimum content as a revolutionary abstraction and then start adding to it degrees of practical existence. There are fascinating details in Badiou's mechanism, but the point is that the truth of an event spread out at intervals into a delayed continuum, forced into a set with the original, can be analyzed formally. The precise form of delayed simultaneity is an open question, but for the moment, I want to consider the sense in which this form is or is not temporal.

Though Badiou posits the 'organized control of time', he elsewhere argues 'against time'.[22] He advocates the 'abolition of the category of time', proposing not even to 'subsume' time under truth or politics, but to subtract time out of the decision of what is to be done, so that politics will decide 'what can be' and not just 'what can come' (46). Yet even here, Badiou retains time-*like* features of events, like dates and sequences (48): dates to cluster phenomena around names (like '1917'), and sequences in a set of fidelity-acts (following 1917 with the Chinese Revolution, May '68, etc.)

In *Logics of Worlds*[23] Badiou says that to be political, a subject crushes the past and sacrifices the present. By 'atemporal filling of the abolished present', the subject becomes 'contemporaneous with the evental present, without necessarily incorporating itself into the present' (61). Again (citing Althusser, 509), events are sequential without being successive: 'the history of a world is nothing but the temporal figure of this universality of its exposition' (321). 'Change' occurs only in the sense that potential variety is built into an event. The actualization of variations does 'constitute the object over time' (359), but the range of changes is determined by the 'transcendental regulation' of its possibilities and the 'transcendental evaluation of intensity', that is, the probability of one change or another based on the degree to which that variation is essential to the phenomena making up the event. To put it strongly, 'true duration can only be that of consequences', a 'distribution or envelopment of intensities'. Badiou interprets the probable consequences of an event in terms of the logical dependence of one variation on another (369). A battle, to use one of his examples, may occur in phases, but these are immanent in the set of phenomena that the battle organizes in its conceptual scheme. The phases may be called 'points', hinged by decisions (401), and take the event in one direction rather than another (decision-points are the 'principles' of politics, 405), but the logical sequencing of event-variations is distinct from what others call the contingencies of history, the genealogies of encounter, or the power of anachronistic event-lag. Badiou certainly derives interesting analyses of battles, political demonstrations, and cultural revolutions, based on category theory and isomorphic set-structures,[24] but it is difficult to accept that contingencies, encounters, and lag could not *also* play roles in political event-making.

Indeed, in *The Century*,[25] Badiou adds a historicist construal of a-historical events. In previous centuries, people may have actually experienced the world in long durations, as a history beyond individual human vision, with meaningful cycles and memorials; but 'today', entering the twenty-first century, 'we have entered a period of a-temporality and instantaneity' (105). Debord makes a similar point about the loss of history in repetitive production,[26] a theme found from Heidegger to Negri. Virilio[27] laments the 'elimination of delay' by the global media's ability to present the whole world at once. (The American FCC regulates that there be a 7-second delay on radio broadcasts, so in fact, 'live' means precisely that much delay.) For Virilio, 'realtime' undermines real time, eliminating the time to decide what the present should become. Adorno and Horkheimer lament the way Hollywood movies promise a story that never comes through, and so has nothing to offer but what is already in the first scene.[28] Badiou is not a lamenter, but he does think we ought to do something about the loss of time. 'If we wish to attain the real of time, we must construct it': not by returning to cycles or explanatory predecessors, but by 'becoming agents of truth procedures' (106). Time is not an illusion, but it is a consequence of decision procedures. If time is infinite, for example, it is because we have decided in favour of infinite sets. For Badiou, time ought to operate in that realm of politics faithful to decisions and their follow-throughs, to cover the sequences we live through. This contrasts with Benjamin and Adorno, for whom anachronism can resist the present with 'shocklike flashes of obsolete elements'.[29] Badiou sees a more programmatic use of the past. But the only way to accomplish a program is to subtract the momentum of the past, subtract the givenness of the present, and subtract the consequentialist values of the future; in short, to decide on behalf of time by subtracting past, present, and future.

Badiou's theory of sequenced non-succession sounds Timaeist. But what does it mean for intervals to be condensed into simultaneities, or for a multiple that occurs at one instant to include subsets that are distanced from each other across time, given that multiple simultaneity is not a Platonic ideal but a real event in the material world? This would have been unthinkable if simultaneity were an instant without temporal difference, and delay were succession without shared temporal reference, that is, if political concepts were universal, and political changes were contextual. Whether or not Badiou's theory of logical unfolding exhausts politics, the model of simultaneously existing multiples whose subsets are distanced across time does describe aspects of revolutions and their faithful retroactions, in which delay is integrated into simultaneity. While a decision made in 1789 does not actually have 1949 in its set of referents, the former is in a sense about the latter—not just because it is forced into being about it later on, but also because decisions are about a virtual time-line whose set it opens. If we are right to append an analysis of decisions with virtual futures to Derrida's to-come, we should also append a virtual world temporal ontology to Badiou's. Badiou's theory is not the *totum simul* of anti-time-passage philosophers, for whom future events are designated simultaneously on a time-line; Badiou's point after all is that the faithful future is not presented in the originary event set. His theory is more like a possible worlds ontology that takes possibility seriously as a distinct realist category. It is like a temporal set that includes one member, along with a procedure for diverging into a set that includes that actual member plus one shadow member. That is what a now is like when a decision occurs. As long as decisions do not occur, the now is regular: its predecessor is its outside influence—actual but external—just as its successor is its probable outcome. But with decision, the now becomes irregular: its pre-evental situations and faithful renewals are possible but internal, given in the moment but without being-there, or in other words, simultaneous but delayed.

Derrida, Balibar, and Badiou describe genuine political phenomena (welcoming refugees, post-national societies, and expansive revolutions), though with disparate variations on delayed simultaneity and simultaneous delay. To choose which temporal structure is the true one would downplay something real in politics—whether the role of waiting, negotiation, or shock value. What keeps political analysis free for construction is sensitivity to the variability of simultaneity and delay-structures and the capacity of these structures to ward off the twin inactions of succession and currency.

It should be possible to mobilize simultaneity and delay in action. And it should be possible to experience, and experiment with, simultaneity and delay in direct sensuous perception.

Chapter 9

Simultaneity and Delay in Music

Theme of the chapter: Simultaneity and delay are made sensible in music. Post-tonal music reveals and constructs novel temporal forms.

Apollinaire's 1914 essay '*Simultanisme*'[1] labels cubism the art of simultaneity. Duchamp in 1915 calls his painting, *La mariée mise à nue par ses célibataires, même*, a 'delay in glass'. Adorno bases his aesthetics of delay on Berg's 1915 *Three Pieces for Orchestra*.

How is the virtual simultaneity across differentials actually experienced? What practice would make actual time run on a multiplicity of virtual rhythms? Composers and performers of music experiment with pure time to discover and create new kinds of simultaneity and delay within perception.

It seems obvious that when tones are played simultaneously, they make a chord; when the same notes are played in succession, they make the melody. But Oliver Sacks describes a musically sophisticated patient who after a brain injury is unable to hear simultaneities as chords.[2] She still hears all the notes in a string quartet, but as separate horizontal voices—all at once, but not as harmony. Sounds at the same time do not make a chord. Sensuous simultaneity needs to make an object *out of* the same-time material. Only the constituted object is fully simultaneous.

Succession too is complicated. In traditional, tonal music, a melody starts on a tone that defines a scale, a tonic. The melody meanders through other notes on the scale, transitioning to other scales, and in the end returns to the tonic where it began. Interesting music stretches the rules, but traditionally, energy comes from succession: it progresses, repeats with variations,[3] and comes home. The anticipation of return also accounts for harmony: when the penultimate note of the scale is played, we anticipate a return to the octave. The chord that returns to the tonic centre is harmonically consonant. For its part, the final note must be prepared for, or it will not feel like a successor.[4] In dance, as Barba and Savarese illustrate, succession is built out of innumerable details of moves and breaks, halts and counter-impulses, balance and resistance.[5] We find it natural that dissonance is resolved in consonance, that themes are developed, and that sonatas and symphonies begin and end in certain ways. However, why music ought to conform to natural expectations, and why it should contract melodies into simultaneous chords or lay harmonies out in succession, is not obvious.[6]

Rhythm is especially complicated. Meter, or beat, is not the same as rhythm.[7] The philosopher Viktor Zuckerkandl points out that most music in world history does not have a regular beat. Polyphonic Western music is the exception, needing a regular beat to coordinate different melodic lines simultaneously. The linguist Émile Benveniste argues that the Greek root of rhythm, *rheo*, refers to the gurgling of a flowing stream, continuous but irregular.[8] Cognitive science finds rhythm and meter in unrelated parts of the brain.[9] Even in terms of metrics, a measure in 4/4 time does not quite mean four equal beats. 4/4 time emphasizes the first note, and secondarily the third, then pulls the fourth towards the beginning of the next measure. Compound meter (mixing duple and triple pulses) requires even more listening power to hear the rhythm. Poetics employs additive metrics, combining different types of syllabic emphasis (stress, vowel length, duration, intonation, word ending, not to mention the rhythmic effects of rhyme and alliteration, and of stanza and longer structural formation). Within a phrase, composers and performers can interject rubato, where a note steals length from its successor. Deleuze says that Schumann's beats express panic, whereas Mozart writes in rhythmic 'microblocks'.[10] Current classical music can be written without measures at all, and of course unmeasured rhythm may still be rhythm, from chant to free jazz.

Even the idea of tempo, or speed, is complicated: tempo may refer to the score or the performance; it may refer to the density of notes per measure or per second; it may refer to changes in speed rather than frequency of events, that is, to accelerations and decelerations that require a short delay for the listener to pick up the thread; it may refer to the cognitive capacities of the listener or to the data-carrying capacity of the acoustic medium; it may refer to possibilities left open for performers to insert notes. The conductor Celibidache emphasizes that if one chord is played too soon after the last, the listener will not hear the richness of either, but if it is played too long after the last, the listener will not hear their unity.[11] Playing a piece at a proper tempo has less to do with speed than with 'richness of expression'. The same holds for comic timing[12] and magic tricks.[13]

The composer Philippe Manoury[14] notes that ambiguous and pluralized tempo has always existed in Western Classical music: Beethoven's Sonata

Op. 110 has four different tempo indications on the first page alone. More recently, Stockhausen's *Klavierstücke 2* is 'hyperstriated', dividing one measure into one section with 5 beats and another with 4, then articulating the 5 into 7/8 and the 4 into 11/12, but with fluid windows within. Boulez often begins a composition with synchronized beats and then allows asynchronous waves to diverge. Asynchronous divergences are less of a problem than subsequent re-synchronization. In temporally complex pieces with many live performers, like Johann Strauss's *Staccato Waltz*, only eye contact with the conductor, not mathematical indications in the score, can lead to synchronized tempo.

(a) Simultaneity and Seriality: Schoenberg and Adorno

Proust calls music 'a bit of time in the pure state'.[15] The thesis that music is an art of pure time, more about time than sound, even that music is the condition of time, may not be as exaggerated as it first seems. At first, it sounds like hyperbole: how could there be music if there were not already time for music to pass in? But if Husserl's account of synthesized time is right, time in general depends on particular time-lines generated from concrete objects. In that case, music (melody, progression, dissonance, blocks of sound), like any field of expression, has to generate time and cannot presuppose it. And music is a sphere in which the virtual differentials within simultaneity become actual. Even so, music is not more the construction of time than any other object-field is. The musical construction of time is not isolated from the way that performers' bodies, buildings' sociality, and scores' histories also construct time at the same time. As Bernard Sève[16] says, music both produces time and is related to the time of life.

I am not committed to any one school of composition or theory. I use a range of resources to cut a pathway through the multi-layered simultaneities and delays that make music a structurable bit of pure time.

The composer Elliot Carter wrote in 1965[17] that current music had yet to concentrate on experiments with time. Carter experiments with 'parquet floor' rhythms, 'themes stated simultaneously—interacting, yet stratified by different tempos'. He scores specific metronome indications for accelerando and ritardando but imagines a piece of music that has no steady tempos at all or that mixes fixed and unfixed tempos.

Jonathan Kramer[18] calls this the 'postmodern concept of musical time': the 'lack of concern for organic unity, [along with] the use of historically and/or culturally remote sounds' (342). From Beethoven to Ives, temporal forms (successive and simultaneous, coherent and conflicting) put the listener into a state of temporal disunity (and put different listeners into different temporal states).[19]

Two candidates in music might be considered strict simultaneity. The first is the simultaneous sounding of two or more notes: a chord. Later, we will see it is impossible to protect such simultaneity from the infiltration of delay. The second candidate for simultaneity would be the eternal in music. Musical time could be the moving image of musical eternity, whether the eternal is transcendent revelation, or something more mundane like a score that persists after a performance is over. Jean-Jacques Nattiez[20] both shows how the eternal in music (the synchronous and reversible *langue* of Orpheus) combats the temporal (the diachronous and irreversible *parole* of Chronos), and also undermines that dichotomy by showing how poiesis involves author, performer, and listener in a history of responses.

One might look for changeless bits of music in stand-alone phrases or motifs that can be inserted anywhere in a piece. This works in literature too, as Valéry says: 'A *description* is composed of phrases which one can, in general, *invert*: I can describe this room by a sequence whose order is more or less indifferent. The look wanders where it wishes. Nothing is more natural, nothing is more *true*, than this vagabondage, since... *truth is chance*'.[21] Kant defined rooms as simultaneous if we can experience them in reverse order. By analogy, a musical motif that may be placed at will anywhere in the piece is a wandering representative of simultaneity within the succession of musical time.

The Husserlian problematic comes to a head with the difficulty of establishing intentional simultaneities in multi-layered music, and it is even more difficult when music becomes atonal.

Schoenberg's twelve-tone music, beginning around 1905, does not return to the tonic. It gives all twelve semitones in the octave equal weight in every piece, without deciding which is the tonic, that is, without a scale. When designing a composition, the composer begins by arranging the twelve pitches, or the intervals between them, into a freely chosen order, called a row or series. Once the first tone in the row is played in the composition, it must not occur again until all the others have been played in the order designated by the row. The return to the first is delayed as long as possible. The composition consists of varying the row, so the original row may not be apparent to the listener. The composer is allowed to divide the twelve-tone row into chunks: for example, into two groups of six, which can then be stacked simultaneously. This will not produce harmony in the classical aesthetic, but it is harmony in the minimal sense of notes played at the same time. The composer is also allowed formulas for augmenting intervals, or for antagonistic functions, or circular permutations—as long as the tonic does not dominate.

The main variations involve order, and because of this, Adorno says that twelve-tone music is more about rhythm than melody.[22] One can play the row in order, or backwards, in retrograde. One can flip intervals upside down, transposing the second tone to the eleventh and so on, by inversion. One can do both at once, in retrograde inversion. One can do all this while stacking some portions of the rows. One can play parts of the row with different instruments, varying volume and tempo. One can play the row in order with one instrument, and the same row at the same time in retrograde inversion with another instrument, and at the same time two partial stacks in inversion with a different instrument, but slower. It seems at first that a twelve-tone composition will last twelve notes and then stop. But a whole world of beautiful and expressive music opens up.

The 'harmonic problem' in twelve-tone music is a problem of simultaneity. Every composition includes a roughly equal number of all twelve semitones, so the order of tones in the row is what gives a piece of music its distinctive sound. But once the row falls into stacks, most or even all of the tones may sound simultaneously. But then it does not matter which twelve-tone row had been chosen. Music theorists George Perle and Martha Hyde suggest a few solutions.[23] Old-fashioned tonal elements could be reinstated briefly to individuate harmony. Or one could stipulate that only a few tones be stacked at the same time. But these are makeshifts. A third option is what Perle calls 'the verticalization of nonadjacent linear elements' (87). This is paradoxical: verticality means simultaneously played tones, while non-adjacent linearity implies succession.[24][25] The idea is that certain intervals may have the same effect whether they are played as a chord, or in series, or in counterpoint. So when logic calls for playing certain tones simultaneously, but strict simultaneity would confuse the harmony, the simultaneity of the chord can be staggered. The tones will be heard as a simultaneity yet not all at once, as if simultaneous sound events do not even have to be temporally adjacent.

Adorno's thesis is that twelve-tone music is more energetic—even, in a sense, more successional—than traditional classical music, precisely because its simultaneities have more tension. Chords in traditional tonal music are just condensed melodies, not really simultaneities. Conversely, melodies that can be condensed into chords are not really successions. Tonality diminishes both the fullness of the moment and the suspense of delay. In traditional classical music, one starts with a sound, it goes here and there, and in the end retreats, and turns out not to have gone anywhere. Of course, this is unfair. Adorno himself is passionate about classical music, from Bach to Beethoven to Mahler. Nevertheless, circular development is a genuine problem. In contrast, twelve-tone music posits a plurality of notes that do not naturally move through one another, but are juxtaposed in a singular way; we hear pure simultaneity, not condensed succession. And once a chord is independent of melody, the progression from one simultaneity to the next is less predictable, a more dynamic succession.[26] Greater simultaneity thus frees succession.

Adorno's commentary on Berg's *Three Pieces for Orchestra*[27] focuses on the third movement, the 'March'. Traditionally, a march has regular rhythm, fanfare triplets, and other features; in Berg's hands, only 'molecules' of those elements are retained, as 'montage' rather than development (191). This piece is not technically twelve-tone, but has atonal intervals, and unstable themes without privileged development. Without 'recapitulations', 'without looking back' (190), Adorno says, it is able to 'contradict the recurrence of an identical being', and achieve the 'irreversibility of time' (190). What is retained of the march is 'inexorable forward movement' (190), linearity in the best sense, as opposed to cyclical return to the tonic. Paradoxically, the 'symmetry and repetition' of traditional marching tunes make for 'static architecture' (192), so they are not suitable for forward marching.

It seems counter-intuitive to associate the absence of development with forward movement. But this is Adorno's point. Irreversibility is a function of unharmonized simultaneity. It is not really instants, lined up horizontally, that constitute succession. It is rather coexistence and vertical stacking that tumble into horizontal succession. Because each instant is stacked out of irreducible differences, the sequences that link them cannot rely on thematic identity; they can simply proceed along one of the many unprivileged lines in the original stack, with nothing but difference to turn to. An 'irresistible flow' is created by the 'opulent overlaying of voices and constructs' into a 'jungle' of sound, by 'the unfettered increase of simultaneous sounds', a 'constructivist', 'exorbitant plenitude', prefiguring 'vertical infinity', an 'infinitely differentiated sound spectrum' (192–5). 'The simultaneity of countless melodic events is an immediate source of tension… the relations they form with each other point beyond themselves, towards progression and resolution'. Especially when shared across different instruments, 'the de facto coherence of a simultaneity is lifted apart' (195). In short, simultaneity stores up non-unity, and *becomes* succession (193).

Indeed, 'the more that compulsively simultaneous events are presented, the more they strive to expand' (194) into longer and longer compositions.[28] We might say that retrogression is what gives forward motion its time-arrow, distinguishing motion from mere momentum. Berg 'was captivated by the mirage of reversing the passage of time through music', by the retrograde motion of a twelve-tone row (197). Forward motion is pent up, delayed by sudden simultaneity, and we have to catch up with it either by accelerating to where it should have been, or back up to the original row, and then fork forwards along a different past.

This leaves us with a paradox of identity and difference in movement. Traditional music returns to its beginning by resolving it, but of course the end is not exactly the same as the beginning. Most people would call traditional motivated development irreversible. In contrast, atonal music undergoes variations, but can return to the exact twelve-tone row at any point and start fresh. Many would call this non-traditional music a return to the same. Adorno's thinking is that traditional music promotes the feeling in retrospect that the development was inevitable, that the difference belonged to the identity, and that it just took time to lay out its unchanging truth. It is true that twelve-tone music returns to the identical row, but precisely because it cannot provide anticipations with closure, it returns to the same row each time as a different forking possibility; it takes no time at all to lay out endless succession. Now whether this succeeds, whether auditory simultaneity is preserved even when staggered, whether non-adjacency is preserved when verticalized, whether the sound field can bend time in and out of simultaneity, whether simultaneity can be a medium of movement, can only be judged by listening, which the reader should now do.[29]

There are many analyses of simultaneity in post-tonal music. Pierre Shaeffer's 1960s work on 'concrete music'[30] demonstrates the inseparability of pitch, timbre, volume, and attack. Transpose a song to a low enough octave, and we no longer hear it as melody, only as rhythm. Remove the attack, that is, remove the first moment of a tone's duration, and we will not know what instrument played it. Pierre Boulez distinguishes striated and smooth space, pulsed and amorphous time. Karlheinz Stockhausen's essay 'how time passes…' serializes duration in the same way that Schoenberg serializes pitch. Synthesizers calculate polyrhythms that no human performer can play. New graphic systems of notation leave it up to performers to decide whether to play the musical events simultaneously or in succession.[31]

One of the most innovative experiments with simultaneity arrives with 'spectral music' in the 1980s, deploying the harmonic spectrum. When a string vibrates, it produces a primary tone, as well as a series of faint overtones. A C-string produces a C, then a quieter C an octave above, then a quieter G a fifth above, a still quieter C a fourth above that, and so on. The human ear can pick up several overtones. Overtones play a role in differentiating the timbre of instruments. Without them, tones sound synthesized and unnatural. Now, because the second overtone is a fifth above the primary tone, it sounds pleasant when tonal music uses fifths to advance a melody, since the listener already simultaneously hears them subliminally. But of course, music is not obliged to obey acoustic expectations. A composer can use the harmonic spectrum to make in-harmonic sounds, as did spectral composers at the Institute for Research and Coordination in Acoustics and Music (IRCAM), founded in Paris by Boulez in 1970.

If a piano string vibrates beside a violin, the violin's string begins to vibrate at the same frequency, and will produce the same tone, and the same overtones, as the piano. The same conversely, except that the violin can play notes in-between the notes that the piano can play, so some violin tones will not set any corresponding piano string vibrating, and these will not produce piano overtones.[32] In principle, putting different instruments beside each other produces different subsets of overtones, and thus different music, with the exact same score. In practice, most overtones would be too quiet to hear, but they can be amplified. Even better, computers can calculate, or distort, overtone effects, and generate violin-like overtones wherever the composer desires, even if the original note had not been played at all. Computers can create virtual instruments with distinctive harmonic series. In principle, a set of instruments can play just one note in unison, and the multiple overtone series, where each overtone of each instrument feeds back across the orchestra, will yield an enormous plenitude of sound. Now let the composer take this block of sound and sculpt it, subtracting overtones as a sculptor removes parts of a block of marble. In a sense, this will produce melodies, but melody will have less to do with pitch than with timbre. In a sense, the piece will undergo succession, but succession will feel like a simultaneous block presented to the ear one region at a time (the way we see one at a time the simultaneously existing rooms of a house). In fact, time-flow in these pieces is carefully constructed. Gérard Grisey describes his composition *Vortex Temporum* as 'drowning in the vortex of time'. So these sound sculptures are not simultaneities without succession. But simultaneity and succession are not exclusive categories: simultaneity is the differentiated field from which successions are forced out, the virtual coexistence of distant historical events. Simultaneity has a delay-effect.

Tristan Murail's compositions use simulated re-injection loops. From the early days of tape, composers experimented with recording music, then running the recording through a distortion device, then running that distortion through another distortion device, re-injecting the successive results in a loop, sometimes playing all the versions back simultaneously. Murail simulates this mechanical procedure using natural instruments and live performers. One performer plays a tone, then another repeats it after a delay, with various distortions, planned or improvised. The reader will have to listen in order to judge whether she hears delayed simultaneity.

Experiments need to be followed up with transcendental deduction. Does the fact that we can experiment with differential simultaneities in music imply that time in general needs to be understood philosophically according to those structures? Of course, if something can be done with musical time, time includes that possibility. But does the fact that we *can* simultanize a delay, or stretch a simultaneity along several lines at once, mean that all phenomena should be described in similar terms, that perception should be described as dilated impact, that political decisions should be described as possible time-worlds, and so on?

In what follows, I highlight resources for temporal experimentation that have emerged over the last fifty years or so from computer assisted composition. Occasionally, composers use the computer to analyze and synthesize sounds, and then simply lay them down, letting time temporalize the sounds rather than analyzing time itself.[33] In its graphic representation of temporal structure, even an experiment-oriented composition programme like IRCAM's OpenMusic looks similar to programmes like Cubase, which was designed for mixing popular music. To be sure, OpenMusic and the other resources of IRCAM can determine complex durations of musical 'patches'. A phrase can even be determined by a kind of backwards causality, shaped by its not-yet-played successor. These ideas are extremely interesting, but some of them still treat time as a kind of segmentation. Some interesting patterns of time are used in programmes like PureData (an 'unstructured environment for describing data structures'[34]) and Iannix ('a graphical editor of multidimensional and multi-formal scores, a kind of poly-temporal meta-sequencer'[35]). These graphic interfaces with dynamic time imagery treat time more as forces than segments. The reader should try composing something with these programmes to decide whether they are conducive to temporal experimentation.

Even when dealing with computer assisted composition, we have to consider the phenomenology of temporality. For Étienne Darbellay,[36] music reveals 'phenomenologically' how deterministic chaos works, particularly in relation to the simultaneity of what is in the present and its differential prolongations: the 'interface of temporal solidarity'. Musical phrases stretch out by different temporal magnitudes, phenomenologically by 'masks of time' (205) and technically by 'multiple temporal monitors' (336). Thus the present escapes itself: 'Tempo serves as "vanishing points" [or "points of flight"] in the elaboration of this dynamic of the present' (206).

I distinguish two levels of complex simultaneity: relative simultaneity (series that may or may not look simultaneous depending on conditions) and inter-temporal simultaneity (complexities across contemporaneous series), en route to showing how simultaneity becomes delay.

Relative simultaneity

The composer Karlheinz Stockhausen's[37] essay '...how time passes...' studies the parameters of rhythm and melody. If a melody is speeded up enough—a whole Beethoven Symphony played in one second—we hear all the successive notes at once. If one succession of notes is laid over another at a different tempo, we hear the two speeds averaged into one.[38] Simultaneity is thus not a primitive intuition but is relative to parameters: when we are interested in speed, we call tones successive, and when we are interested in information density, we call them simultaneous.

The project of 'total serial music' was to design rows not only for pitch, but also for other parameters like timbre (orchestration) and note duration. Temporal serialism bears on 'tempo (on the large scale) and momentary fluctuations and vibrations of sound (on the small scale)' (88).[39] Just as a tone row lists the twelve semitones in a chosen order, which becomes the 'theme' of a composition, so likewise a composer can make a row of durations. One might choose a row that makes the first note in a composition a 1/2 note, the second a 1/16 note, the third a double whole note, and so on.

However, problems arise for serialized duration. (a) Once durational rows are stacked simultaneously, they tend to be heard as one loud regular pulse. Iannis Xenakis's alternative[40] is that if simultaneous time-strands produce nothing but a mass of notes, it is an opportunity for stochastic music to progress by statistical variation rather than serial phrasing. (b) Using all the note-durations from 1/32 note to double whole note with equal frequency makes the average duration of notes too long to be interesting. The challenge in serializing time is to 'maintain different tempos simultaneously' (90), as an 'elastic play of time-strands' (92), without letting the different tempos blend into a single mass of sound with no audible tempo at all. One of Stockhausen's solutions (*Gruppen*, 1955–57) is for two orchestras, placed on opposite sides of the audience, to hurl musical material back and forth distinctly yet simultaneously over the listeners' heads. Another solution (*Zeitmasse*, 1955–56) calls for musicians to play 'as fast as possible', to keep distinct tempos unreconciled, both as structure and as spectacle. Morton Feldman and Hughues Dufour decelerate their chord progressions or arpeggios to such a slow pace that the listener forgets what the previous one was, undermining the anticipations and fulfilments that constitute the very possibility of succession. John Cage's *As Slowly as Possible*, 1985, is currently being performed by an electric organ (which can hold each note for an extremely long time) in Halberstadt, Germany; begun in 2001, it is scheduled for completion after 639 years, in 2640. (c) The parameters of pitch and duration can be hard to keep distinct (89). If a string vibrates at 32Hz, it is perceived as pitch, but if it vibrates at 4Hz, only its duration is heard. A series of sounds at one range of frequency is melody; at another range, it is rhythm.

In 'Time and Interval', the composer Bernd Alois Zimmermann[41] aims to explain how a sound aggregate with an objectively null time interval might be experienced as a succession. Different levels of the same experience can have different temporal events as their intentional objects, a thesis shared by Husserl and Webern. Indeed, there are endless subjective means of synchronizing and contemporizing time-lines. Just as Goethe envisaged flashing a lantern across a statue to make it appear to move, we might drive by an orchestra pit, hearing just a moment of music go by, with a Doppler effect pluralizing the simultaneity effect.

Sometimes the relativity of succession and simultaneity is achieved by technical means, and sometimes by manipulation of phenomenological receptivity. A case of the former is Darius Milhaud's piece from the 1940s designed to be either two string quartets or one octet, depending on whether performers choose to play them successively or simultaneously. A case of the latter is Debussy's *Pelléas et Mélisande.* Debussy says[42]: 'I wanted the action never to stop, for it to continue uninterrupted… In hearing a work, the spectator is used to experiencing two very distinct sorts of emotions: the musical emotion on the one hand, the emotion of the character on the other; generally, he feels them successively. I have tried to make these two emotions perfectly blended and simultaneous. Melody, if I can say this, is anti-lyrical. It is powerless to translate the mobility of souls and of life. It is essentially suited to the song, which confirms a fixed sentiment. I have never let my music *hurry up*, or *delay* (*brusquat ou retardat*), by technical means, the movements of the sentiments and passions of my characters'. Composing music and drama simultaneously is precisely what allows them separate rhythms. If they are successive, they deprive each other of their prolongations. But if they are simultaneous, one of them can withdraw and still be quasi-present, and let the other come to the fore. In short, simultaneity allows things to take their time gradually, without speed, whereas succession requires fixed starts and stops. Paradoxically, simultaneity is gradual and succession is segmented. Succession plus simultaneity, according to Debussy, avoids delay. But if continuous motion without speed avoids delaying the passions, it does after all prolong the instant into the continuity, delaying its presence for the duration of the action.

The association of simultaneity with continuity rather than with instantaneity, or in other words, the alliance rather than the conflict of simultaneity and succession, is also essential to impressionism's contemporary rival: futurism. Moholy-Nagy writes:

> *vision in motion*
> is simultaneous grasp... It instantaneously integrates and transmutes single elements into a coherent whole...
>
> *vision in motion*
> is seeing while moving.[43]

Inter-temporal simultaneity

The simultaneity of two tones is relative to the capacity of the listening organs and to tempo, complexity, and pitch. When two standards are applied at the same time, the same sounds can be both simultaneous and successive at once. Simultaneity can be simultaneous with succession. I call that inter-temporal simultaneity.

Apollinaire's 1914 essay '*Simultanisme-librettisme*'[44] theorizes about the simultaneous poetry readings practiced by Dadaists, futurists, and cubists of the period. The essay challenges Henri-Martin Barzun's *Manifeste sur le simultanisme poétique*. Apollinaire says that Barzun's vulgar notion of simultaneous poetry is satisfied with several voices speaking at once (or a phonograph playing voices simultaneous with a live reader). The problem is that voices read poems one word after another, successively. On a page, words are written simultaneously, but the question is how to liberate the spoken word from succession. Many have tried: Villiers de l'Isle-Adam, Marinetti,[45] Duchamp (the champion of both simultanism and delay), and

Picabia, not to mention Mallarmé's 'typographic simultaneity'. Of course, if voices speak exactly the same phrase in unison, it would be nothing but a chorus. Apollinaire's simultanism is meant to be textually polyphonic, not just to stagger the same text in a canon like *Frère Jacques*. Though the words are still read one after another (Apollinaire does not anticipate a Stockhausen style reading of all the words in one instant), the listener should 'immediately conceive the simultaneity of what they express', in a kind of 'ambient lyricism' (890). Upon listening to a good simultaneity poem (such as '*L'amiral Cherche Une Maison à Louer*' (1916) by Tristan Tzara, Richard Huelsenbeck and Marcel Janco[46]), I find I do get a sense of simultaneous expression, though I get it better on second listening, which somewhat defeats the purpose.

Though Apollinaire prefers oral to written simultaneity, he admires Blaise Cendrars and Sonia Delaunay's written poem, *La Prose du transsibérien et de la petite Jehanne de France*, 1913, whose typography uses colour contrast to help the reader apprehend several texts at once. This is analogous to seeing several figures on a poster, or to the way 'a conductor reads at one blow the superimposed notes in a score' (890). Apollinaire calls cubism the art of simultaneity, 'representing several faces of figures and objects at the same time' (891), but he thinks futurism offers even better paradigms. In futurist paintings, the implied speed of human or industrial gestures piles facets upon facets (e.g. Giacomo Balla's *Sculptural Construction of Noise and Speed*, 1914–15); whereas in cubism, simultaneity owes less to speed than to the plurality of positions, and hence to the dismemberment and mutual intrusions of part objects: 'the gift of ubiquity'.

Neither oral nor written performances are strictly simultaneous in the absence of succession, either in production or reception. The best solution is to simultaneously and synaesthetically combine art forms in such a way that the simultaneities in each compensate, relatively, for succession in the others. The Delaunays are self-conscious heroes of simultaneity, working simultaneity into plastic arts, literature, and music (891). Sonia Delaunay designed and wore a 'simultaneity dress' to the Bal Bullier dancehall in Paris in 1914.[47] One wonders what simultaneity behaviour befits simultaneity clothing.

Inter-temporal simultaneity differentiates coinciding successions. Grisey[48] calls his *Vortex Temporum* (1995), 'an application of the same material to different times'. The same sound appears simultaneously in 'ordinary time, more or less inflated (*dilaté*) time, and more or less contracted time' (5). Arpeggios are restated at several speeds simultaneously in the first and third movements, and slow steeply in the second. Silent interludes between movements create 'a time of waiting, a bridge thrown up between the time of the listener and the time of the work'. Grisey makes the Bergsonian point that different species follow different 'time constants', for example, that the 'times of language and respiration' for humans, and times of sleep for whales, are like overlying times in a piece of music (6). His claim that time-inflation creates the 'sensation of spherical and vertiginous movement' (6) might be hyperbolic. But he is not exaggerating when saying that the work 'is the history of an arpeggio in space and in time' (7). Gérard Zinsstag, who wrote the CD liner notes, says it 'penetrates a different temporal dimension, where material is abolished in favour of duration' (4).

Catherine Basset's[49] analysis of gamelan music arrives at a similar conception of mixed-rate succession. Gamelan consists of several percussion instruments, treated as a single instrument with many performers. Notes are distributed among the performers, according to how many are present. Each one plays 'unequal and syncopated rhythms', but the result is 'a single uninterrupted line of strokes of equal duration' (101). 'The work is chopped into small complementary tasks, simultaneous and non hierarchical' (102). The division of labour makes one and the same rhythm both staggered and simultaneous. Meanwhile, gamelan mythology invokes the battle against linear time, *kala* the evil time demon (110). To escape time, humans both avoid the moments of conjunction when things happen, and also reiterate time cyclically, falsifying time's claims to linearity, turning time against itself into a mere symbol of time (111). Calendars in gamelan cultures do not have single 'dates', but situate each event on many 'superimposed' cycles of differing lengths, all turning 'simultaneously' (112). If a cosmology of beats eschews measurement of length, accent, and regularity, should we still call it 'rhythm' (133)?

Pierre Sauvanet's[50] volumes on rhythm in cosmology and phenomenological ontology analyze the metaphors that associate rhythm with life and cosmos: that rhythm is the pulse and heartbeat of life, or the harmony of the spheres, or the geometry of the cosmos, or the poetry of language. Sauvanet is sceptical. A lot of rhythm is nothing like pulse; the physical cosmos is not harmonious; poetry needs more than rhythm.

Sauvanet grounds temporality in polyrhythm without metaphor. He begins with Husserl's point that rhythm is an experience not of regular progression, but of multiplicity, 'at the same time simultaneous and successive' (102), combining 'lacunas and polyrhythms'. If an ego-substrate existed, it would be able to hold multiple time-lines together. But since phenomenology refutes the substrate, time has to be unified by construction, by rhythmic processes themselves. Rhythms are therefore the 'becoming' of time (103, 108). Rhythm is 'neither in time nor is it time itself' (104). It structures the possibility of time, so it cannot be a particular content of this or that segment of time; 'rhythm *temporalizes*' (105). It gives 'the sense of time' (105), in three senses of 'sense' (in French): 'sensation, orientation, and signification' (106). Rhythm renders time sensible; it assigns a (spiral) 'arrow' to time (which would otherwise be 'an empty, non-oriented form'); and it interprets temporal phenomena (106). We can even say that 'rhythm plays "*against*" time', in that it plays with the flux of time and reconstitutes it (108). The point is that time is ordered by rhythmic intervals, discontinuities, schemas of movement and silences (112–3). Continuous, homogeneous time would not really be time at all (115). Of course, it is not just in sound, but also in visual and haptic spheres (caress, sculpture, dance, even the taste of wine), that 'time becomes *palpable*' (123, 151).

Rhythm is discontinuous (otherwise it would not mark passages), as well as unified (since time is nothing other than the arrows that multiple rhythms produce). Polyrhythm feeds questions and answers back in a 'virtuous circle' (130–1). Citing Valéry, it is '"*as if all the terms were simultaneous and current*, but appeared only successively"... It appears to me in diachrony, but I only grasp it in synchrony... Rhythm thus makes two dimensions coexist' (131). Furthermore, rhythm cannot distinguish private from intersubjective experience. Quoting Valéry again: '"You sing, I march according to your song. You cry, I suffer"' (136). We could add, with Hegel, that the vibrations that each perceiver receives from the rhythms of the other (the composer, the performer, the musical instrument, the air) is also the listener's return-to-self, her recollection both of embodiment and of abstract thought.[51]

Three developments in computer-assisted composition can illustrate the cognitive transition from relative to inter-temporal simultaneity.

José-Manuel Berenguer[52] derives rhythm from our tendency to divide 'packets' of information, and store them in short term memory. 'The illusion of synchrony depends on the decisive capacity of the device', particularly 'electronic delay devices, and even more so with reverberation' (29). For example, handclapping appears as a discrete sound only because we packet it into a simultaneity. If consciousness makes several packets out of the same information, we hear it diachronically. Conversely, a single sound appears successive if modulated by reverberation algorithms with delay circuits. (Delay before reverb begins is called 'pre-delay'.) Berenguer may or may not be right about cognitive packeting, but he is right to challenge the dichotomy between continuous succession and holistic synchrony. Electronic and digital devices manipulate the packaging of musical information, and so manipulate simultaneity-succession relations. Time becomes an experimental plastic.[53]

Leo Küpper[54] makes a similar suggestion. Whereas analogue music creates the illusion of a linear time-arrow, digital music works with 'granular quanta' of time (96). Instead of 'packets', Küpper speaks of the 'thickness' of musically discrete instants, where some length or combination of sine waves creates a 'latency' effect, like a 'halo' (99), that in turn creates the feeling of time stopping between one musical effect and the next. (Analogue music has similar 'superclock' effects (94).)

Finally, Curtis Roads'[55] premise is that most 'sound', namely 'zones of intensity and frequency', is inaudible to the human ear. Once computers organize sound-particles at various frequency scales to allow the ear to pick up what it normally misses, 'using sonic brushes to scatter them on the canvas of time' (153), we may not only change our consciousness of synchrony and diachrony, but our whole sense of sound.

In sum, categories like packeting, thickening, and scattering open up the description of rhythm, and of simultaneity-succession models beyond the normal scope of 'rhythm', in a way that is both artificial and yet free of previously assumed technical constraints. On the down side, Philippe Schoeller[56] complains that computers have two features that prevent them from achieving the temporal complexity they promise. First, a digital programme is indifferent to the complex time of the performance and indeed pre-exists, or at best exists, '**SIMULTANEOUSLY**' with it (204, bold and caps in the original), thereby vitiating the temporal dynamic. Second, a programme can only define temporal complexity in binary moments, as 'successive micro-simultaneities' that are temporally much less complex than the potential of traditional instruments (206). One might reply that the binarity of the programming does not entail binarity in the audible music it produces. Even so, the simultaneity of computer programming and musical performance is not just the simultaneity of matter and expression but is insistently made visible in musical performance on stage. We have to consider the 'realtime' simultaneity of computer programming with instrumental performance.

The term 'realtime' is sometimes overused to refer to images recorded and broadcast simultaneously (as in live news reporting), but it would be clearer to call that phenomenon 'direct time' (*temps direct*), or 'live'. To refer to images recorded at one time and played back later, we could say 'delayed time' (*temps différé*). We reserve the term '*realtime*' for interactions between a human user and a computer programme, where the human's intervention in the programme and/or the programme's response to human input results in an (almost) instantaneous result in the music generated. 'Realtime' will not simply mean 'present' but 'interactive' in producing dynamic changes across systems with minimal delay. Interaction across media in real time is a paradigm of simultaneity, and preprogramming for future performance is a paradigm of delay—both are interactive.

Inter-temporal simultaneity is a matter of mixed media. The classical music theorist Georgiades Thrasybulos[57] analyzes the origins of polyphonic simultaneity in Western music as the interaction of literary and musical media, beginning with Southern European ninth-century Carolingian settings of the mass. The basic Kyrie is sung with the final tone lasting throughout ('bourdon accompaniment'), simultaneous with an embellished 'interpretative paraphrase'. Polyphony is a consequence of 'polytextuality'; mixing musical phrases simultaneously arises from mixing scripture with exegesis (16). In contrast, Northern European settings of the mass favoured sonority over textuality, and so had less need for simultaneity or polyphony at the level of sound. Nevertheless, these two 'seemingly irreconcilable' models of the music-language relation generated a history of variants on the relation of simultaneity and succession (31). Thrasybulos offers speculative comparisons: improvisation over a figured base as Leibnizian substance-subjectivity; operatic action as Kantian process-subjectivity (109); Palestrina as pre-Kantian thrusts and kicks, non-organic 'pure crystals' of movement (101).

We can schematize three levels of mixed media simultaneity: parallel simultaneity, open simultaneity, and two-era delayed simultaneity. In the first level, a plurality of voices is simultaneously on stage, whether by doubling instruments in unison, or 'padding voices' in opera, which symbolizes the simultaneous juxtaposition and obliteration of individuals in bourgeois society, as Adorno says.[58] Opera, especially after Verdi, does not just put drama to music, it stages several singers with distinct songs at the same time. Without the musical differentiation of singers, the textual differences among songs would be hard to hear. This extends Nietzsche's observation in *The Birth of Tragedy* that ancient Greek drama moved from having one personage on stage to two to three or more. In film, Howard Hawks' screwball comedies introduce multiple characters speaking at once. Performance and conceptual drama frequently overlay voices, as in Janet Cardif's 'Whispering Room': you are directed into a small simulated movie theatre and given headphones; while you watch a (strange) film, you hear on your headphones supposed fellow audience members interrupting the film with irritating chatter. Already in 1927 (the year of Abel Gance's split-screen *Napoléon*), Moholy-Nagy[59] imagined 'Simultaneous or Poly-cinema', which would simultaneously project two or more films about related characters on a large screen, moving each film across the screen horizontally, vertically, or diagonally, so that the films cross at various points, and may or may not merge into a single film when they do. Just as skilled drivers can keep track of many relevant sounds and sights at once, so skilled spectators should follow several letterboxed films simultaneously. It is not just individual voices that can be simultaneous, but whole ensembles of simultaneous and successive voices.

The second level, open simultaneity, concerns not sounds heard, but performances chosen. Extensive research has of course been done into group improvisation in jazz, a field too big for the present chapter.[60] The choreographer Frédérique Belivacqua[61] creates works of improvisation where one dancer is shown on video while another interacts with that image. He wants improvisations to be synchronized, but finds that synchronization is compromised when live dancers are told which gestures to use. It works better when they are habituated not to precise gestures but to intervals between gestures, open 'time windows', and allowed to skip gestures in order to re-synchronize.

In the third level, two-era delayed simultaneity, one of the contemporaneous ensembles has a temporal structure that lags culturally behind the other. Adorno[62] says that, 'Wagner is the first to insert the uneven development of the arts' into a single work, a historically staggered simultaneity also found in cinema.[63] 'The human ear has not adapted itself to the bourgeois rational and, ultimately, highly industrialized order as readily as the eye, which has become accustomed to conceiving reality as made up of separate things, commodities, objects that can be modified by practical activity. Ordinary listening, as compared to seeing, is 'archaic'; it has not kept pace with technological progress' (13). Harmony and rhythm today (not counting new music, which Adorno and Eisler promote to Hollywood) are still tied to the collective church aesthetics of community response, whereas the eye has moved on to free particulars. To be sure, collective synthesis can be progressive in art and politics, but the fact that the ear is still unable to mark the modernist difference between the individual and the collective makes it open to 'misuse', 'retrogression and confusion', to making capitalist values appear as if they were invented by and conducive to individuality. It is as if two distinct eras exist simultaneously, received by structurally and politically distinct sense organs that happen to function on the same head at the same time. Adorno is not saying that non-contemporaneity is progressive on its own (indeed, he wants the ear to catch up), the way Deleuze does when advocating the visual/auditory staggering of the cinema of Straub and Huillet, but he does describe what non-parallel simultaneity looks like.

Inter-temporal simultaneity thus leads us to describe simultaneity by non-simultaneity. If simultaneity is dissonance, non-simultaneity is delay.

(b) Delay and the Unpulsed Time-Arrow: Boulez, Deleuze and Spectral Music

Not all post-classical composers favour simultaneity. Wolfgang Rihm[64] argues that simultaneity produces only exterior variations. Without predecessors, novelty is banal, research without discovery. Interesting mutation, he thinks, operates by succession and makes a difference to what it returns to (174). Rihm's music is certainly not classically tonal, but shares with it the art of 'waiting', of the 'after-effect'. Still, a post-tonal composer cannot avoid the fact that something other than tonal continuity has to hold successive elements together. If that job cannot be done by thematic development, it has to be done with more purely temporal 'synchronization' (177). It is this temporal logic—not mere succession, but succession held in place by re-established simultaneity—and not mere simultaneity, but simultaneity that mutates—that I call delay. Just as simultaneity in music almost immediately shows its relativity, its inter-simultaneity, and its delay-structure, so delay will take us to the inter-delayed contemporaneity of different time-arrows, and hence to a simultaneity structure within delay.

We said that simultaneity is found in a chord, succession in melody and chord progression, and delay in the process of resolving dissonance. In fact, we find delay in music at all levels: in material acoustics, in echoes and reverberations; in harmonics; in cadences; and in purposive listening. Delay sometimes means waiting for the end of a symphony that takes a long time to arrive. Sometimes it means a beginning that does not stop sounding, stuttering like a too-repetitive motif, or games of start-and-stop. Delay covers equally the not-yet and the still-going.

It is common to define all art forms that rely on anticipation, memory-constructed narrative, and a desire to find out what will happen, in terms of delay. The essayist Pascal Quignard[65] writes: 'To introduce delay into immediacy: that is music (or memory), and that is why *mnemosyne* and *musica* are the same'. Roland Barthes[66] writes that classical narrative literature is all about delayed discovery. Deleuze,[67] writing about detective fiction, says that, 'Literature is like consciousness; it always delays'. Of course, not all classics are like detective fiction, eventually answering questions posed at the outset. Post-classical music illustrates a distinctive kind of delay. When Schoenberg[68] says that 'The construction of a basic set of twelve tones derives from the intention to postpone the repetition of every tone as long as possible', he introduces a kind of delay that is not later resolvable into temporal consonance, but a delay that is musical for its own sake.

We can use François Decarsin (who cites Deleuze) to sketch a threefold distinction between simultaneity-music, succession-music, and delay-music. (i) Pure simultaneity-music occurs in the pure present, like acoustic ornamentation (94). Sequences in Schoenberg, for example, while complex, bifurcating, and superimposed, show no evolution, development, prolongation, causal projection, repetition, or stabilization (98–100. This is not Adorno's view.) Its unpredictability (106) concentrates listening on the instant without a future (108), 'simultaneity' without 'return' (109). In a similar way, spectral music creates non-chromatic time, a qualitative difference of consecutive instants without causality, appreciation without memory, re-absorption and entropy without order (111). The pure presence of the experience need not involve shock or emptiness; repetition can produce the same effect of non-succession (112).

(ii) Succession-music with a past-future arrow is defined by the aim of return (119). The paradox is that the return undermines the time-arrow, resolving into changelessness, reversibility, redundancy and the halt of information, suspension of the future, circularity without new beginnings (122). To avoid this, Boulez uses retrograde motion to create futuricity inside 'isochrony', interference between past and future (124).

(iii) Murail's '*Désagregation du temps*' offers another way to abandon development without abandoning temporality: temporal de-temporalization, or delay-music. Murail's 'rarification' of musical information, which noticeably increases over time, creates a sense of intemporality and even eternal return (138). Similarly, accelerating from a slow start can generate trance-like ecstasy out of monotony (148), a state from which exit is delayed.

Juxtaposing these three strategies, we find an ambiguity in the notion of delay oriented music. As music focused neither on a simultaneous present nor on a developmental succession, it can refer either to music whose temporality is amorphous, unpulsed, or smooth; or to music that incorporates specific but non-holistic time-arrow fragments. For my taste, the latter is richer. But the attempts by many composers to generate a time-arrow out of entropy indicate the former, namely, a procedure for breaking down the accordion of projection and return. This is one plausible definition of 'smooth time'.

Boulez's *Penser la musique aujourd'hui*[59] introduces 'smooth time' into the discussion of tempo. Tempo is relative. It defines speed not absolutely, but within 'boundaries of a chronometric field' (51). Tempo varies along several parameters. Just as pitch can be changed not only punctually but also glissando, so tempos and durations can be changed 'diagonally', so that at no moment in the sliding scale is the tempo fixed. Tempos can be undirected, floating, or undefined. In part, this depends on psycho-perceptual factors. In 'striated' or 'pulsed' musical space or time, following a law of 'identity by reproduction', the perceiver hears divisions (pitch, duration, timbre, or volume) as 'landmarks' (52, 86–90). In contrast, in 'smooth' or 'amorphous' time, the perceiver feels the freedom, sometimes disorienting, to make divisions at will, as if the sounds themselves had no speed or temperament (85). Smooth time is not without events; there are musical materials to divide, but no one of them is a reliable landmark. Smooth pitch distributions appear more orderly to people with perfect pitch, but no one has perfect durational sense, so landmark-free time feels smooth to everyone. If landmark striations are too dense, they may feel smooth, and if events in smooth time are too rare, they may feel striated. Nevertheless, the overall structure should clarify whether time is smooth or striated.

It is not just a local series that can be striated or smooth. A longer time span, a 'block of duration' (55), 'envelope', or 'time bubble' (58) can have a degree of temporal density, a fixed or mobile 'index of distribution' (67) of regularities and irregularities (52). For that matter, a symmetrical relation can transpose into an asymmetrical one, and vice versa, changing rhythms around an axis. Many distinctions are opened up by this analysis: 'straight vs. curved time', according to whether tempos evolve or repeat; 'focalized vs. unfocalized time', according to whether there are one or many axes of symmetry; 'homogeneous vs. heterophonic (polyvalent) morphologies', according to the distribution of striated and smooth segments (92–3, 105).

By structuring the structureless, Boulez carves smooth time. Smooth does not mean shapeless or vague; it takes a patient accumulation of detail to make rhythmic material smooth. Sauvanet[70] considers what it might mean for musical time to be *strictly* 'arrhythmic' (139). It could mean (1) non-rhythmic or foggy; (2) under-rhythmic, as in irregular heartbeats; or (3) over-rhythmic, as in sympathetic vibrations that bring down bridges and neural networks (140–41). But none of these arrhythmia is really smooth. In contrast, Sauvanet calls 'eurrhythmia', 'a "healthy" association of different rhythms' (145): not a single iso-rhythm, but rhythmed rhythms, an organized rhythm that includes a 'normal amount of chaos… leading to a certain state of equilibrium' (145). If smooth time truly meant arrhythmic, rather than eurrhythmic, either it would be impossible, or else it would not have the innovative value it has for Boulez and Deleuze.

At a conference at IRCAM in 1978, Boulez invited Deleuze, Barthes, the composer Luciano Berio, and others, to comment on five pieces of music, each illustrating a temporal structure: pieces by Olivier Messaien to illustrate time rhythm, Georgi Ligeti to illustrate time structure, Stockhausen to illustrate time fields (determined by performer's breathing speed), Boulez to illustrate time objects (based on resonance or decision), and Carter to illustrate time layers (accelerations and retards). Deleuze comments that all five illustrate non-pulsed time, and thereby make pure duration audible. He defines non-pulsed musical time as non-coinciding, non-homogenous, non-communicating sounds. Consciousness cannot unify such sounds. Their only unity exists as populations of oscillators.

Deleuze gives three characterizations of non-pulsed, 'floating', 'variable pulsation': time liberated from measure, individuation liberated from subjectivity, and material liberated from form.

The first liberates smooth time from both regular and irregular measure. Deleuze is not talking about chaotic or arrhythmically empty or neutral time. On the contrary, just as we do not walk or swim by measures, so music 'puts into presence a multiplicity of durations that are heterochronic, qualitative, non-coinciding, and non-communicating'. Deleuze's term 'heterochrony' suits what I call inter-temporality. By analogy,

> biologists, when they study the vital rhythms of 24-hours, do not try to articulate them by a common measure, no matter how complex, or by a sequence of processes, but invoke what they call a population of molecular oscillations, which when coupled ensure the communication of rhythms or the trans-rhythmical.

It is not only electronic music that can oscillate instead of pulse. Messiaen's palindrome rhythms oscillate on natural instruments. The point is that the aim of non-pulsed music is not to produce muddy or obscure rhythm but to lay down rhythmic elements that retain distinctive patterns even when they couple.

The second characteristic of non-pulsed time individuates phrases and rhythms without form or subjectivity. Programme music associates a musical phrase with a landscape, a persona, or an epoch. In contrast, non-pulsed music creates landscape, persona, or rhythm, inside the sonority of time itself. It is difficult to pinpoint Deleuze's intent here, partly because the view that musical phrases are associated with real people, places, and times already seems hyperbolic.

The third characteristic liberates matter from form. Deleuze rejects musical form, since matter is already 'elaborated' into differential forces. The point again is not that non-pulsed time is amorphous, but rather that it is elaborated internally and not externally. Deleuze's twist is not just to replace form with force (as Aristotle did). It is that matter is 'given the task of rendering sonorous and audible those forces which are not, by themselves, audible'. Among the forces that are not in themselves audible, but which music can make audible by unleashing matter, are time, duration, and rhythm (as well as non-temporal forces like intensity). Smooth or non-pulsed does not mean non-articulated, but rather means elaborated without formula, indeed to be so elaborate that no description of its form could help the ear listen to it more easily. 'Smoothness' may be a misleading term for this. For nothing is more rhythmically, chromatically, and materially jagged than smooth time, liberated from formal striation. So if the smooth is not muddy, what is it?

In '*Le systeme et l'idée*',[71] Boulez associates smooth time with the collapse of the time-arrow. One cannot 'verify' a theme if it is merely 'superimposed' on material without motivation (71). But 'delayed' verification (90) is only a problem for a theme made up of unambiguous tones. Once tone is regarded as a cut-up effect, a passing slice cut out of a vast system of audible sound, even accidental sound-productions are part of its system. Boulez can thus say that a fragmentary part is heard 'simultaneous' with the whole (93). The theme is no longer a basis for deducing tones in succession, but is now just one fragment among many drawn from a virtual whole. There is still a system, but no temporal orientation.

In short, smooth time has micro-structures but no time-arrow. It allows for local simultaneities and delays, without universal simultaneity across the sound field, and without global delay across the full duration of the piece.

Jérôme Baillet[72] defines time-arrow music, like Kant, by its irreversibility. Non-arrowed music may have an order but will appear not to (165). Of course, even if the performer is given free choice in ordering the elements, a performance will still pass by in some order. Different performances can reverse the order, but to create music that is reversible on a single playing, the composer has to resort to special devices. One extreme form of non-arrow music is one-note music, for example, Yves Klein's imagined 'Monotone Symphony' (169) or Giaconto Scelsi's 'Four Pieces for a Single Note' (170). If the whole piece of music is monotone, it does not matter which part we hear first. But in most cases, so-called non-directional music is, more accurately, music whose moments are unconnected (172).

Non-directionality and unconnectedness are sometimes equated with unpredictability. But predictability is not the key point. Steve Reich's de-phasing of repetitive cycles that are continuous and predictable, are nevertheless slow enough to lose the sense of direction (179). Such effects generate the sense of non-directionality, without literally being non-directional. In other words, smoothness can be generated as an effect without entirely abandoning the time-arrow.

In fact, Baillet sees a return to a time-arrow starting with the later music of Ligeti, which rejects classical repetition, symmetry, ostinati, and periodics, yet aims for a fragile new version of direction (194). Similarly, Murail's echoes and 'the universe of reverberation', or degrees of 'delay' (198–9), aim for 'temporal immobility' that is nevertheless temporal. Still, the directionality of a segment in non-arrow music is due more to the field character of the whole than the anticipatory quality of the segment that precedes it.

Vagueness or ambiguity can in some cases be what de-motivates an ordered sequence. But vagueness is a passive way of de-motivating succession; nothing prevents a listener, performer, or composer from experiencing a motive why one segment follows from the last. To keep the time-arrow from controlling the music, to keep smooth time from striating, and to keep simultaneity and delay audible in themselves, the music itself must be non-entropic. Musical segments must not have endpoints at which their force runs out of steam. Smoothness must be maintained in continuous robust motion, reproducing itself in successive incarnations: either in a positive form that reuses musical segments even after they have been fulfilled or in a negative form that suspends fulfilment.

Positive de-arrowing gives a musical element not only actual content, but also fictional or virtual content that will never have been entirely heard, and whose continuation remains unexhausted. Costin Cazabin[73] invents a musical application of three-valued logic, in which propositions can be counted true, false, or 'fictional' (or 'diversified'). 'True' musical propositions are time-arrowed, but their 'fictional' truth-values are not. Cabazan thinks this allows 'backwards causality' in musical time-lines (306) as well as 'temporal evasions' (311). If tones have fictional counterparts, the explicit melody might run to an end, but its fictional counter-melodies run on.

Negative de-arrowing nullifies the consequence of a sequence even when the consequent evidently occurs (like the retention of unfulfilled anticipation even after it has been fulfilled). Suspense music for film makes an interesting case. Adorno and Eisler[74] emphasize the paradox: suspense music intensifies the expectation of surprise, but by pre-figuring the violence to come, it nullifies the surprise (10). Its time-arrow telegraphs the content of its minimally delayed sequel, and so it loses suspense and undercuts delay. To be fair, musical elements in film have to be short. Even a two-minute development of a theme feels endless if the visual content remains constant during that amount of time. 'Motion picture music cannot "wait"' (63). It has to move without delay to the next climax: 'like a *stretta* without a preceding *piu moderato*' (66). Indeed, Hollywood prides itself on subordinating suspense to continuity. Continuity editing—match cuts on action, the 180 degree rule of composition, point-of-view transitions, motivated sound tracks—is designed to make transitions between shots seamless and invisible, even though the techniques are striated by codes, signifiers, and perceptual habits. The typical Hollywood time-arrow is thus at once exciting and boring, programmatic without relevance, directional without advancement, purpose without purposiveness. Of all the regressive effects of clichéd tune-making in film, the worst is that, like so much in capitalism, it is objectively boring; yet the 'psycho-technique of the studios deprives the consumers of the awareness of the boredom they experience' (84). In short, the arrow goes nowhere.

The alternative to anticipatory suspense would be a violent scene that unexpectedly interrupts continuity. Paired with scenes that 'jump forward or backward in time' (20), atonal music can play an effective role. By expressing drama in the moment, music is politically liberating, teaching our perceptual faculties how to reconcile socially predictable processes with unprepared singularities. Interjecting song and dance can also interrupt continuity in a liberating way. The non-arrow liberates the temporal order.

Paradoxically, whereas striated time is continuous, smooth time is jerky. Jump cuts free the editor's control of time from the demands of the narrative object, and hence make the passage of time visible for its own sake. By not giving its sequel away, time-jumping jump-cutting music treats its sequel as genuinely not present until it arrives, leaving the sequel undecided, and genuinely delaying its teleological power. Paradoxically, the future is possible only if there is no time-arrow (or, if one prefers, a time-arrow is possible only on condition of maximum delay). To be sure, fast jumps are not the only way to suspend suspense; it can also take the minimalist or gradualist form, as Hughues Dufourt's[75] ultra-slow chords de-emphasize the need for resolution: 'generalized suspension, perpetually deferred (*différée*) resolution' (8). The twelve-tone composer Meyer Kupferman's languid score in Alan Baron's noir *Blast of Silence* (1961) is a case in point. Whether by slow or fast techniques, the arrow is defeated when tightly defined delay-objects are loosened into general delay.

This does not mean that suspenseful music does not exist. It may indeed be a transcendental universal that all music, and all time, has anticipatory, time-arrow force. But there is an opposing force as well: the temporal jump. To activate smooth time in a milieu that is at the same time striated, smoothing mechanisms must be engaged. One level of any piece of music is directional, so to smooth it out, a second has to redirect it. This leads us to the problem of secondary iteration, or reproduction. The concealed partial and broken arrows in recorded music reveal the final point about delay in musical time.

Bernard Stiegler analyzes technological reproduction as a delay that conditions originality.[76] Charlie Parker's youthful improvisation to phonograph records might be called, in today's 'multiplied time', the technological production of memory. The material in the recording he plays back is obviously not being performed in the present. But the playback presents the original music as 'deferred time' in real time (130). This is a kind of delay, but since many people own copies of the recording, it is 'simultaneously reproduced in several places at once' (133). Delay of playback permits the simultaneity of subsequent innovations: the simultaneity of the improviser with the recorded music, and the simultaneity of different improvisers playing along with different phonographs in different cities on the same evening. It is always relevant to ask what one is doing while receiving a transmission: driving away from the Martians while listening to the radio, attending a demonstration while watching it on an iPhone, grading papers while watching a film noir DVD.[77]

For Stiegler, delay is indispensable for authentic temporality, creative production, and political recognition. The apparently instantaneous transmission of images on internet news sites, for example, undermines real time just because it does not build in a 'delay' (137). It conceals the fact that it takes time for one event to affect another, to have meaning through after-effects, in the time lag between inception and reception. And it conceals the possibility of autonomous deliberation and responsible intervention. Presenting today's news merely as today's is not presenting it as news of the world at all, but only as an image, not as real time but as vanishing present. To read an event, as opposed to merely having it pass before one's perceptual field, is to grasp it 'at the same time' as one is aware of reading it (138), to grasp the time at the same time as a second time. The delay in grasping it is needed in order to make the viewer simultaneous with the event's reality. Loss of delay undermines the time character of the event, since it relaxes the tension between the current event and the process that makes it history (139). Citing Virilio (whose *Empty Sky*[78] makes similar points), Stiegler says that TV creates a 'delay of the delay' (149). The industrial market economy is the age of false realtime, without memory, hallucinating 'humanity without time' (156). In authentic history, events would be 'put in reserve', 'delayed in principle' (150). The structure of delay—its authentic relation to reproductive simultaneity on the one hand and its inauthentic relation to the illusion of immediacy on the other—varies with the chosen technology: each has its own brand of 'delayed time' (*temps différé*) (153).

Stiegler's generalization that computers eliminate after-thoughts may be exaggerated. But the principle that after-thoughts allow events to be improvised, offers a way to define smooth time doubled up by delay.

There are many cases where music and other arts converge by technological delay. Antoine Hervé[79] says Cage's preparation of pianos manipulates sound in 'delayed time'. Sound editing technologies like the old 'Rhythmograms'[80] synchronize sound with motion pictures. The photographic image is paradigmatic delay, captured by 'time delay mirrors' (*miroirs de retardement*), freezing a moment in the development process.[81] A photo keeps past time, demobilizing a scene, painting in the medium of delay, fixing its elements in frozen simultaneity.

Reception adds a final converging series. Jean-Noël Portugal, a designer of interactive video games, distinguishes three levels of narrative temporality, of which only two are found in traditional music. Music contains one measure of time indicated in the score (the time narrated in the narrative), and a second measure enacted in the performance and experienced by the listener (the time it takes to narrate the narrative). In literature too, of course, the reader can choose to read slowly or quickly; a viewer can pause a DVD on a single frame. As Hugo Münsterberg says, film frees mental time from physical time.[82] The extra sequencing layer in interactive games is that they will be played (not just viewed, practiced, rehearsed, repeated, re-ordered, or improvised upon) for different amounts of time per day. The game may be repeated for improved results, or to find new obstacles, add new characters, or compete with different opponents. Success for a game writer is measured by how long users play: 'runtime'. Game-writers do not know what they have made until they hand it over to players; nor do online composers who make their music and programming available free to users to play with.

Art that the audience contributes is a key part of Duchamp's formula.[83] Of his *La mariée mise à nue par ses célibataires, meme,* he writes:

> Sort of subtitle: *DELAY (RETARD) IN GLASS*: To use 'delay' instead of picture or painting; painting on glass becomes delay in glass–but delay in glass does not mean picture on glass–

> This is simply a means of coming to no longer consider that the thing in question is a picture—to turn it into a delay in the most general possible way, not so much in the different senses in which delay can be taken, but rather in their undecided meeting. 'Delay'—a delay in glass, as one would say a poem in prose or a spitoon in silver. (41)

The painting covers two plates of glass with a sort of dividing line. The upper plate holds the bride and the Milky Way. (When we say what the picture represents, this is in the dada sense: as Tzara says, everything is dada and dada is everything.) The bottom plate holds the bachelors in uniform (a milkman, a soldier...), a mill generating love gasoline, filters, a chocolate grinder (bachelors grind their own chocolate), something that chomps the gas into little pieces where they circulate and get picked up by the Kodak lens, and turned into something more airy. Then the gravity manager pulls them up, and they cross the boundary into the bride's area. The bride has her own way of working with the love gases, with the female pendulum, the wasp, and the cylinders that expel cinematic outflow into the Milky Way that drips back down to the bachelors. The bride and the bachelors work on separate machinist schemes, with a little point where something of the one goes to the other. It is possible that the bachelors never get their bride. Or maybe the bride is a prostitute seducing the bachelors. Or maybe it is all about the bachelor's getting rid of their uniforms and becoming nude and rarefied, or becoming-woman. Or maybe it is more about light than matter. The '*mar*' in '*mariée*' may be Marcel, a self-portrait.[84]

In terms of content, the bachelors never quite get their bride; their consummation is delayed. In terms of materiality, the painted sheet of glass appears to be a running machine, but of course it is a motionless visual presentation, a frozen movement like Duchamp's *Nude Descending a Staircase* (1912), as frozen as Laocoon. Perhaps we should not interpret Duchamp's remark about delay at all, since dada means nothing and culture is stupid. Dada titles have nothing to do, as Max Ernst's series, '*La femme cent têtes*', equally 'means', 'The woman with a hundred heads', and sounds like, 'the woman with no head' (*la femme sans tête*). Duchamp's *Green Box*, now published as a book but originally sold as a box full of scrap notes about the Large Glass, including the fragment above concerning delay, may fall into this category of unintelligibility. When dada shows you a thing, just look, in the same way that when dada tells you to go to a dada dentist, you might as well. This itself is a delay tactic on the way to language.

Duchamp's painting is not random, but hyper-rational. Duchamp invented his own science of colour, and techniques of keeping different manufacturers' pigments from fading over time on a glass surface. He experimented for weeks with scatter patterns of paint dust. The *Green Box* is stuffed with intricate procedures and diagrams for aleatory production. In like manner, dada and surrealist 'ready-mades' were made 'ready' by elaborate schemas preparing for spontaneity. Tell yourself that on Thursday you will wander aimlessly around a flee market, and at precisely 4:00 in the afternoon, turn 90 degrees to the right, and call the first thing you see a 'ready made' work of art. The art will be chosen by a time lag that defines instantaneity: 'canned chance' (*hasard en conserve*, Suquet, 37).

It is possible to see the machine in Duchamp's Large Glass as a series of ready-mades. Intricate technical procedures derive measurements for its images by repeated accident. When Duchamp says that, 'delay in glass does not mean picture on glass. It is a delay in glass as you would say a poem in prose or a spittoon in silver', he means not only that the bachelor-machine does not accomplish its goal. More important, delay is the medium of art, like a poem in prose. What is the delay made out of? Glass. In several passages (e.g. *Duchamp du signe*, 163), Duchamp says that the bride, the 'hanged female, is a "projection" of an "imaginary entity" in four dimensions in our world of three dimensions, and in the case of the flat Glass, a re-projection of these three dimensions onto a surface in two dimensions'. These pataphysical calculations may, as for Alfred Jarry,[85] constitute pure speculation on the law of universal exception. Every painting, and indeed every object in three or fewer dimensions, delays by omission its temporal life. But Duchamp is not so much generalizing, as putting the Large Glass forward as an exception. The desiring machine of the bride, materially and poetically, is a delay mechanism in essence, and the painted glass rushed, after a mere eight years, into existence, before the thing itself even became real. The delay is merely in glass, not in reality. A poem in prose could be rewritten in verse, as a spittoon in silver could be recast in bronze; the delay could have been made out of canvas. But the point is not that something other than glass could have been the material (though the transparency of glass makes it a suitable support for the blank background). The point is that what is in glass is not a painting but a delay. Duchamp is not a painter, but a delayer. The art form in question is not line-drawing or colouring or composition, it is the art of time-restraint, the art of keeping things from their temporal processes, the art of transposing the most dynamic substances (desiring machines) into situations where there are not enough dimensions for that dynamic to unfold. Why would an artist want to do such a thing? There is little choice. Time gets delayed whenever there is an unmarried subject who desires. Since every delay eventually gets enacted, and wasted, the artist tries as long as possible to suspend it in glass.

Simultaneity and delay in music and the other arts are not parameters that limit what can or should be done but fields of experimentation with time as the sensible realm. At this point, the reader will have to conduct her own experiments with instruments and ears. The fact that we can produce staggered simultaneities in music suggests that in perception, epistemology, politics, and even daily life, we might profitably learn to stack our simultaneities with more temporal dissonance.

Chapter 10

Conclusion

(a) Organizing the Simultaneity-Delay Timescape

Theme of the chapter: In its simplest form, simultaneity consists of two or more events at one time, and delay consists of one event at two or more times. By varying the formal relations between the unity and multiplicity of events, and the unity and multiplicity of time, we might deduce forms of simultaneity and delay. Formal systematization does not substitute for concrete ontological and phenomenological description, but it can generate conceptual experiments and organize the simultaneity-delay timescape.

I have spoken of staggered simultaneities in rhythms and overlapping series; of delayed reactions, leftovers, time-lapses on and off schedule, finite, infinite, and infinitesimal deferrals, stages, relays, and a-periodic intervals. What simultaneity and delay share concerns the shifting contours of an event along different temporal series playing out over the same duration, or in different phases connected across durations separated by distance. It takes converging multiple time-lines to constitute simultaneity, and it takes splitting singular time to constitute delay. A full moment of time exists only when there are many time-lines. It takes two times to make one time, and one to make two. Simultaneity and delay are mutually constitutive structures that together generate, and problematize, succession.

We might say simply that simultaneity consists of two events at one time, and delay consists of one divided event at two times. But different kinds of ones and twos characterize different kinds of delays in relation to different kinds of contemporaneities.

For example, if we think of a single event occurring twice—once temporally and once eternally—in the city of God and in the city of man—then the whole realm of succession is a delay of the eternal realm (the eternal is already real and actuality has to catch up with it). But if we make a slight change, and think of the event occurring once actually and once virtually, then not only is the virtual now delayed in relation to succession (since the virtual is what else the actual event can become later), but multiple variants in the realm of succession now appear as equally real and contemporaneous series. Or instead of pairing a quasi-temporal mode of the event with the temporal mode, we could directly divide the duration of an event into segments (with overlapping borders and intervals between significant transformations) and think of an event occurring across different blocks of time. When an event is repeated on another occasion (like an annual celebration), we might say that the same event is prolonged even though the time of the event is plural. Alternatively, when an event is segmented, cliff-hangered, and renewed after an interval (like a romance left and picked up years later), we might say that the time of the event is prolonged even though the event is plural. Some sequences occur over two segments of time, some over two occasions of time. We might think of deceleration as delayed oneness and acceleration as condensed twoness. Temporal fractioning, which sometimes problematizes the counting procedure, allows for crossover events, prematurities and anachronisms, and (sometimes helpfully) inconsistent matters for decision. $N + 1$ and $N - 1$ temporal inductions, numerical calendars, set theoretic relations, pataphysical mathematics, and all sorts of one-two relations each reveal something about simultaneous and delayed events across time.

Forms are not guaranteed to have instances and, without concrete analyses, do not ensure an ontology of time. But abstract deduction from principles, whether in the manner of Proclus, Spinoza, Hegel, or Badiou, generates options for novel structuration, for the discovery and inventory of new sorts of events, for the invention of new philosophical problems, and for new temporal practices.

A general theory of multiplicity might thus entail forms of simultaneity and delay. The following suggestions are not complete or axiomatic. Still, a formal combinatory might be traced from ways in which phenomena are at once one, two, multiple, infinite, and uncountable.

The principle of the combinatory is that simultaneity and delay involve superimposition, identification, coincidence, equation, or unicity as well as difference, multiplicity, variation, or divergence.

The elements are time-segments and events.

The variations range over instants and flows.

Recursively, levels of simultaneity and delay exist in staggered simultaneity.

The criterion depends on whether we define events by beginning-points and endpoints (quantitatively distinct and temporally contiguous) or by meaning (in the way a war can take place on-and-off, without contiguity and with only rough quantitative distinctness). In the former case, events will be simultaneous if they match points on a timescale; in the latter, if they pertain to the same situation. In the former, immediate causality may test for simultaneity; in the latter, indirect causal influence is an equally good test.

The postulate is that it is possible to measure whether two events occur at one time, whether one event occurs over two times, and so on.

The antinomy is that if two allegedly simultaneous events are in all ways exactly one in time, then they are the same identical event, not two simultaneous events. But if they differ, then they will not entirely overlap in time and will not be strictly simultaneous either.

Limiting ourselves to events and times, units and dyads, simultaneities identified out of successions, successions identified out of simultaneities, and simultaneities identified out of other simultaneities, we can lay out, among others, the following structural possibilities that determine corresponding phenomena:

1. One event in one time. Zeno's paradox of the arrow implies that if only one event can happen at a time, time is impossible. For events to give place to successors, they cannot exist at independent times, they must in part coexist at the same time. Hence,
2. Two events in one time. This is the normal case of coincidence, whether by instant or by duration, found respectively in chords and polyphonic melodies. Things are more complicated when durations unfold from instants, as when melodic variations are played out from a single chord. Subsequent variations may count as simultaneous even when their rhythms are out of synchronization, as long as they were based on simultaneous starting-points. In short,
3. Two events at one multiple time (i.e. in coexisting time-lines). A complex of distinct rhythms may undergo non-isomorphic but contemporaneous changes of pattern. Marshall McLuhan, cited by John Cage, says that these days, 'Everything happens at once'.[1] But when temporally differentiated events overlap, there can also be,
4. One event in two times. Stoics posit cycles in which the history of the universe is repeated identically over and over; the same event occurs at two times, but at the same time in each cycle. The idea of 'two times' here is ambiguous, but similar ambiguities are also found in everyday events that continue from one moment to the next. When an event includes some features that are new and some that continue, as in technologies that are simultaneously cutting-edge and traditional, it is not just that one phenomenon has several temporal phases but that an

innovation occurs, disappears from history, then perhaps after a time lapse of centuries, picks up and carries on. These are cases of,

5. One multiple event at two times: delayed reactions. A single event occurs in separate parts, at different times. Causality at a distance may count as simultaneity at a distance. There may be multiple time-signatures within a single duration, namely as,
6. Two times in one time. Prophecy telescopically compresses present and future. In secular imagery, many times might literally be visible as one, as in cinematic sheets of time stationed from foreground to background, exhibiting different times of the action. The fact that there is no way to synchronize clocks across different space-time frames of reference shows that the same events occur at irreducibly different times at, as it were, the same time. But perhaps we should not speak of 'one time' at all.
7. Two times in one event. Certain historical events happen twice (first as tragedy, then as farce); some must reach their effect in order to have happened at all (perlocutionary linguistic acts); some must have their later stages happen before their earlier stages (faster-than-light subatomic events); some are produced only once they are reproduced (capital accumulation, self-consciousness, inter-textuality).
8. One *or* two events in one *or* two times. Bergson's virtual memory, like Dennett's decentralized brain, exists disjunctively. If the brain never decides whether it does or does not have a given thought at a given moment, the event remains one *or* two. It may always be possible to deduce one more set of simultaneous objects from simultaneities already existing.

The reader will find many other cases, particularly once three and more events are deployed: cases where two simultaneous actual events combine into a third possibility that is delayed; cases where one event is the delayed reaction of two different causes; cases of perception where the difference between simultaneity and delay is indiscernible; cases where rhythms overlap, so that beats from a plurality of them occur simultaneously, but never one beat from all of them (like rhythmic family resemblance, or Bertrand Russell's 'compresence'[2]); cases of simultaneous events that diverge over time or after repetition.

When counting times and events, we are not limited to small rational numbers. We can imagine complexes of events and divisions of time that are numbered by complex fractions, irrational numbers, infinite sets, and power sets. Infinite power set events may be distributed in infinitely simultaneous ways over an infinite set of delayed time-lines.

There is no single structure of time. But all structures of time are structured by succession, simultaneity, and delay. Who can say how many structures of time are yet to be found or made? The theory of simultaneity and delay is the derivation of all such possibilities.

Notes

Chapter 1

[1] Georges Didi-Huberman surveys the problem of anachronism in the history of philosophy—its dangers, inevitability, and value. *Devant le temps: Histoire de l'art et anachronisme des images*. Paris: Éditions de Minuit, 2000, pp. 9–55.

[2] Daniel Rosenberg and Anthony Grafton. *Cartographies of Time: A History of the Timeline*. New York: Princeton Architectural Press, 2010.

[3] Blain Brown, *Motion Picture and Video Lighting* (Second edition). Burlington, MA: Focal Press, 2008, pp. 136–37.

[4] Marc Augé, *Le temps en ruines*. Paris: Éditions Galilée, 2003, p. 39.

[5] Hans-Georg Gadamer, *Truth and Method* (translated by Joel Weinsheimer and Donald G. Marshall). New York: Continuum, 1994, p. 86.

[6] Adolf Grünbaum, 'The status of temporal becoming'. In *The Philosophy of Time: A Collection of Essays* (edited by Richard M. Gale). London: Macmillan, 1968, pp. 322–54.

[7] Gadamer, 'Plato's unwritten dialectic', in *Dialogue and Dialectic* (translated by P. Christopher Smith). New Haven: Yale University Press, 1980.

[8] Edmund Husserl, *Cartesian Meditations* (translated by Dorion Cairns). The Hague: Martinus Nijhoff, 1973.

[9] Ingeborg Bachmann, *Simultan* (translated by Mary Fran Gilbert under the title *Three Paths to the Lake: Stories*). New York: Holmes and Meier, 1997.

[10] André Breton, *Les vases communicants*. Paris: Gallimard, 1955.

[11] Lampert, *Deleuze and Guattari's Philosophy of History*. London: Continuum, 2006, chapter 6.

[12] Rosenberg and Grafton, pp. 197–99.

[13] Helga Nowotny, *Time* (translated by Neville Plaice). Cambridge: Polity Press and Blackwell Publishers Ltd., 1994, especially chapter 1, 'The illusion of simultaneity'.

[14] Gotthold Ephraim Lessing, *Laocoon: An Essay on the Limits of Painting and Poetry* (translated by Edward Allen McCormick). Baltimore, MD: The Johns Hopkins University Press, 1962.

[15] See Fredric Jameson's *The Prison-House of Language*. Princeton, NJ: Princeton University Press, 1972.

[16] Theodor Adorno, 'Berg's discoveries in compositional technique', in *Quasi una fantasia: Essays on Modern Music* (translated by Rodney Livingstone). London: Verso, 1992, pp. 179–200.

[17] Karlheinz Stockhausen, '....how time passes.....' *Die Reihe*, 1956, pp. 10–40. Pierre Boulez, *Penser la musique aujourd'hui*. Paris: Éditions Gonthier, 1963.

[18] Edmund Husserl, *Zur Phänomenologie des Inneren Zeitbewusstseins. Husserliana X.* Haag: Martinus Nijhoff, 1966, s. 38 pp. 76–79; also *Beilage V and VII,* pp. 109–11 and 115–6.

[19] Jacques Derrida, *Speech and Phenomena* (translated by David B. Allison). Evanston: Northwestern University Press, 1973.

[20] Deleuze seems to draw an opposite conclusion regarding simultaneity and spatial totality: 'Time ... has the strange power to affirm simultaneously the pieces that do not make a whole in space...'. Hegel's argument that simultaneity requires that space be totalized, and Deleuze's argument that it not be, agree on the neeed for transportation in any direction whatsoever. Gilles Deleuze, *Proust et les signes.* Paris: Quadrige/Presses Universitaires de France, 1996, p. 157.

[21] Henri Bergson, *Matière et mémoire.* Paris: Presses Universitaires de France, 1959.

[22] Gilles Deleuze, *Différence et répétition.* Paris: Presses Universitaires de France, 1968, pp. 108–10.

[23] Jean-Paul Sartre, '*La fin de la guerre*', in *Situations, III: lendemains de guerre.* Paris: Éditions Gallimard, 1949, p. 63.

[24] Jean Baudrillard, *The Illusion of the End* (translated by Chris Turner). Stanford, CA: Stanford University Press, 1972, p. 115.

[25] For different approaches, see Éric Alliez, *Capital Times* (translated by Georges Van Den Abbeele). Minneapolis, MN: University of Minnesota Press, 1996; Marc Shell, *Money, Language, and Thought: Literary and Philosophic Economies from the Medieval to the Modern Era.* Baltimore, MD: The Johns Hopkins University Press, 1993; Margreta de Grazia, 'Teleology, delay, and the "old mole"', *Shakespeare Quarterly,* 50, 1999, pp. 251–67.

[26] Howard Richards, 'Deference'. *Ethics,* 1964, 74(2), pp. 135–42.

[27] Rem Koolhaas and Bruce Mau, *S, M, L, XL.* New York: The Monacelli Press, 1995, p. 286, lexicon entry under 'Delay'.

[28] Wikipedia, 'Delay Differential Equation'.

[29] Jason Robinson comments on an early version of this chapter. 'Simultaneity and delay: Lampert's already and not yet of now', in Jay Lampert and Jason Robinson (eds), *Philosophical Apprenticeships: Contemporary Continental Philosophy in Canada.* Ottawa: University of Ottawa Press, 2009, pp. 1–11.

Chapter 2

[1] Edmund Husserl, *Zur Phänomenologie des Inneren Zeitbewusstseins. Husserliana X.* Haag: Martinus Nijhoff, 1966, s. 38, pp. 76–79; also *Beilage V and VII,* pp. 109–111 and 115–6.

[2] Retention is a middle ground between succession and simultaneity. In s. 7, Husserl discusses a problem developed by Herbart and Lotze, Brentano and Stern: does the recognition of a past moment in the present require that the past be held in consciousness *simultaneously* with the present moment (Brentano), or is it enough that consciousness pay *continuous* attention to successive experiences (Stern)? Husserl's theory is closer to the latter. Retention locates the past in the temporal mode of having-passed. Izchak Miller calls this the 'Principle of Simultaneous

Awareness', in *Husserl, Perception, and Temporal Awareness*. Cambridge, MA: MIT Press, 1984, pp. 164–5.

[3] Sokolowski argues against my type of reading. For him, the *content* that appears over time may be hard to describe as a series of simultaneous situations, but the *frame* of time-flow is not hard to describe in that way. Robert Sokolowski, *Husserlian Meditations*. Evanston, IL: Northwestern University Press, 1974, pp. 134, 158.

[4] Kockelmans names the three dimensions of time as before, after, and simultaneous, rather than before, after, and present. Joseph Kockelmans, *A First Introduction to Husserl's Phenomenology*. Pittsburgh, PA: Duquesne University Press, 1967, p. 182.

[5] Husserl's melody examples vacillate between themes of cause and effect and themes of impressional quality. The same ambiguity over whether melody involves change in sensory quality, or development of a theme, can be found in Kant. Kant's intensive magnitudes govern continuous degrees of qualities like pitch (allowing for gaps) from 0-to-1 on a spectrum. But this would not account for melodies with deliberately non-continuous intervals and non-obvious resolutions. These require a causal conception of where tones lead, not just a qualitative sensation of tones increasing and decreasing. This is why Husserl's examples of melody deal with objective intentionalities, not just primal sensations. Husserl's examples from music involve tunes with at least a few notes, whereas Kant's sonorous intensive magnitudes make music sound like whistles blowing.

[6] Husserl acknowledges that even the single note C can be broken into different parts of the sound, so even this is not perfect simultaneity but 'simultaneous succession' (s. 41).

[7] Wilhelm Wundt, *Outlines of Psychology* (translated by Charles Hubbard Judd). London: Williams and Norgate, 1902, p. 160.

[8] Soren Kierkegaard, *Either/Or* V. 1 (translated by David F. Swenson). New York, NY: Anchor Books, 1959, 'The Unhappiest Man', pp. 215–28.

[9] Maurice Merleau-Ponty, *Phénoménologie de la perception*. Paris: Éditions Gallimard, 1945.

[10] Deleuze makes the same distinction relative to Bergson: memory images are not prolonged but 'delayed reactions' (*réactions retardées*). *Cinema 1* (translated by Hugh Tomlinson and Barbara Habberjam). Minneapolis, MN: University of Minnesota Press, 1986, pp. 90–91.

[11] Merleau-Ponty begins with a different fourfold, citing Claudel: '*Le temps est le sens de la vie (sens: comme on dit le sens d'un cours d'eau, le sens d'une phrase, le sens d'une étoffe, le sens de l'odorat)*'. The translator wisely does not translate this passage, but I will: 'Time is the sense of life (sense: as one speaks of the direction of a flow of water, the meaning of a sentence, the feel of a piece of fabric, the sense-organ of smell)'. We might connect time-series with direction, time-fullness with meaning, time-order with feel, and time-scope with organ.

[12] See also Heidegger on boredom and 'time dragging', in *The Fundamental Concepts of Metaphysics* (translated by William McNeill and Nicholas Walker). Bloomington, IN: Indiana University Press, 1995.

[13] Bernard Waldenfels, 'Time lag: Motifs for a phenomenology of the experience of time', *Research in Phenomenology*, 30, (1), 2000, pp. 107–119.

[14] Bruce Baugh argues, with Sartre, that delay allows us to hear the difference between others' voices and our own, an encounter across 'diachronic alienation'. *French Hegel, From Surrealism to Postmodernism.* New York, NY: Routedge, 2003, pp. 106–109.

[15] Heidegger, *Being and Time* (translated by John Macquarrie and Edward Robinson). New York, NY: Harper and Row, 1962.

[16] Françoise Dastur says that the 'event of thinking' forgets anticipations and backward references (81) for the sake of 'simultaneity by which two movements of inverted sense subsequently make only one' (10). *Dire le temps.* Fougères: Encre Marine, 1994.

[17] How long things take is ethically complicated. See Karen Houle, 'Emendation, *or* when have we been', In *Lyric Ecology: An Appreciation of the Work of Jan Zwicky,* edited by Mark Dickinson and Clare Goulet. Toronto: Cormorant Press, 2010, pp. 219–30.

[18] For Adorno, insomnia is the 'bad eternity of time', mocking those projects that claim to 'gather past and future into the present'. *Minima Moralia* (translated by E. F. N. Jeffcott). London: Verso, 1978, p. 165.

Chapter 3

[1] Plato, *Timaeus* (translated by R. G. Bury). The Loeb Classical Library. Cambridge, MA: Harvard University Press, 1981.

[2] Plotinus, 'On eternity and time' (translated by A. H. Armstrong). The Loeb Classical Library. Cambridge, MA: Harvard University Press, 1967.

[3] See Werner Beierwaltes, *Platonismus und Idealismus.* Frankfort: V. Klostermann, 1972.

[4] Pasquale Porro, 'Angelic measures: *Aevum* and discrete time', in Porro (ed.), *The Medieval Concepts of Time.* Leiden: Brill, 2001, pp. 131–59. Thanks to Antonio Calcagno for this reference.

[5] Patricia Parker, 'Dilation and delay: Renaissance matrices'. *Poetics Today,* 1984, 5, pp. 514–35.

[6] Thomas A. Carlson, 'Consuming desire's deferral: A theological shadow in the culture of image'. *Parallax,* 1999, 5, pp. 39–55.

[7] Jean Ladrière, '*Raison et eschatology*'. *Revue de l'Université d'Ottawa/University of Ottawa Quarterly,* 1985, 55, pp. 173–91.

[8] Plato, *Philebus* (translated by R. Hackforth), in *The Collected Dialogues of Plato* (edited by Edith Hamilton and Huntington Cairns). Princeton, NJ: Princeton University Press, 1973, pp. 1086–150.

[9] Aristotle, *Physics* (translated by R. P. Hardie and R. K. Gaye), in *The Basic Works of Aristotle* (edited by Richard McKeon). New York, NY: Random House, 1941, pp. 218–394.

[10] Badiou's *Number and Numbers* (translated by Robin Mackey) has an excellent survey and theory of relational numbers. Cambridge, MA: Polity Press, 2008.

[11] Giorgio Agamben, *Le temps qui reste* (translated by Judith Revel). Paris: Éditions Payot & Rivages, 2000.

[12] Derrida, 1999–2001 lectures to be published as *Peine de mort* (edited by Geoffrey Bennington, Marc Crépon, and Thomas Dutoit). Paris: Galilée, 2012.

[13] For ontological and hermeneutical implications, see my 'Origen on time'. *Laval théologique et philosophique*, 1996, 52(3), pp. 649–64.

[14] Tertullian: 'We pray for the delay of the end [*pro mora finis*]', quoted in Agamben, p. 185.

[15] Aristotle, *Nicomachean Ethics* (translated by Martin Ostwald). New Jersey: Prentice Hall, 1962.

Chapter 4

[1] Immanuel Kant, *Critique of Pure Reason* (translated by Werner S. Pluhar). Indianapolis, IN: Hackett Publishing Company, 1996.

[2] Lenin takes '*esse est percipi*' this way: 'Berkeley tries to connect the notion of reality with the simultaneous perception of the same sensations by many people'. V. I. Lenin, *Materialism and Empirio-Criticism: Critical Comments on a Reactionary Philosophy* (translator not named). New York, NY: International Publishers Co., 1972, p. 23.

[3] David Hume, *A Treatise of Human Nature* (edited by L. A. Selby-Bigge). London: Oxford University Press, 1973, p. 76. Graham Harman explains the 'buffering or delay' between cause and effect by positing the unfathomability of objects. *Guerrilla Metaphysics: Phenomenology and the Carpentry of Things.* Chicago, IL: Open Court Publishing Company, 2005, pp. 208–9.

[4] Penelope Mackie considers, sceptically, whether delays can be causes. When April rain delays the usual May forest fires, does delay 'cause' fires to occur in June instead? 'Causing, delaying, and hastening: Do rains cause fires?'. *Mind*, 1992, 101, (403), pp. 483–500.

[5] Relativity physics may undermine Kant's causal account of simultaneity. See Adrian Bardon, 'Kant and the conventionality of simultaneity'. *British Journal for the History of Philosophy*, 2010, 18, pp. 845–56. However, simultaneity for Kant is less about shared instantaneity than about coordinating subjective schemata.

[6] 'Every substance... must contain within itself the causality of certain determinations in the other substance and simultaneously must contain within itself the effects of the other substance's causality... if their simultaneity is to be cognized in any possible experience' (A212–3/B259).

[7] Hegel, *Hegel's Philosophy of Nature* (translated by Michael John Petry). London: Allen and Unwin, 1970.

[8] Deleuze draws the opposite conclusion regarding simultaneity and spatial totality: 'Time... has the strange power to affirm simultaneously the pieces that do not make a whole in space...'. Hegel and Deleuze disagree over whether simultaneous space is totalized, but they agree that simultaneity implies free transportation. Gilles Deleuze, *Proust et les signes.* Paris: Presses Universitaires de France, 1996, p. 157.

[9] Johann Gottlieb Fichte, *Science of Knowledge* (translated by Peter Heath and John Lachs). New York: Appleton-Century-Crofts, 1970.

[10] Friedrich Schelling, *System of Transcendental Idealism (1800)* (translated by Peter Heath). Charlottesville, VA: University Press of Virginia, 1978, p. 111.

[11] Fichte, *Science of Rights* (translated by A. E. Kroeger). London: Routledge and Kegan Paul, 1970. See my 'Locke, Fichte and Hegel on the right to property', in Michael Baur and John Russon (eds), *Hegel and the Tradition.* Toronto, ON: University of Toronto Press, 1997, pp. 40–73.

[12] See Margreta de Grazia on Hegel, Marx, and Derrida, on the (Shakespearean) motif of the mole that burrows underground throughout history until it suddenly bursts forth as modernity. 'Teleology, delay, and the "old mole"', *op. cit.*

[13] Ancient Egyptian painting uses body-twisting devices. H. A. Groenwegen-Frankfort, *Arrest and Movement: Space and Time in the Art of the Ancient Near East.* Cambridge, MA: Harvard University Press, 1987.

[14] Gottfried Ephraim Lessing, *Laocoon: An Essay on the Limits of Painting and Poetry, op. cit.*

[15] Johann Wolfgang von Goethe, 'Upon the Laocoon' (translated by Samuel Grey Ward), in *Goethe's Literary Essays* (edited by J. E. Springard). Whitefish, MT: Kessinger Publishing, 2007, p. 27.

[16] Lessing, 'On how the Ancients represented death' (translated by E. C. Beasley and Hellen Zimmern), in *Laokoon and How the Ancients Represented Death.* London: G. Bell and Sons, Ltd., 1914.

[17] J. J. Winckelmann, *Reflections on the Imitation of Greek Works in Painting and Sculpture* (translated by Elfriede Heyer and Roger C. Newton). La Salle: Open Court, 1987.

[18] 'Lingering' (*verweilen*) over the history of poetry is a theme of Lessing's *Hamburg Dramaturgy* (translated by Victor Lange). Mineola, NY: Dover Publications, 1962, ss. 101–4.

Chapter 5

[1] Max Jammer, *Concepts of Simultaneity: From Antiquity to Einstein and Beyond.* Baltimore, MD: The Johns Hopkins University Press, 2006.

[2] Peter Kosso, *Appearance and Reality: An Introduction to the Philosophy of Physics.* New York: Oxford University Press, 1998, pp. 31–61.

[3] Albert Einstein, *Relativity* (translated by Robert Lawson). New York: Three Rivers Press, 1961.

[4] Hans Reichenbach, *The Philosophy of Space and Time* (translated by Maria Reichenbach and John Freund). New York: Dover Publications Inc., 1958, pp. 143–6.

[5] Vesselin Petkov, 'Simultaneity, conventionality and existence'. *British Journal for the Philosphy of Science*, 40, 1989: 69–76, pp. 70–3.

[6] Alfred North Whitehead, *Science and the Modern World.* New York: The Macmillan Company, 1925, p. 176.

[7] Oskar Becker, 'Contributions towards the phenomenological foundation of geometry and its physical applications' (1923), in *Phenomenology and the Natural Sciences* (edited by Joseph Kockelmans and Theodore Kisiel). Chicago, IL: Northwestern University Press, 1970, pp. 117–41.

[8] Gaston Bachelard, 'The Philosophic dialectic of the concepts of relativity', in Kockelmans and Kisiel, *op. cit.*, pp. 571–2.

[9] Gaston Bachelard, *L'intuition de l'instant* (1931). Paris: Éditions Stock, 1992, pp. 30–1.

[10] Jacques Maritain, '*De la métaphysique des physiciens, ou de la simultanéité selon Einstein*'. *Oeuvres de Jacques Maritain,* Volume 3. Paris: Éditions Saint-Paul, 1984, pp. 237–301.

[11] Richard P. Feynman, *QED: The Strange Theory of Light and Matter.* Princeton, NJ: Princeton University Press, 2006.

[12] John McTaggart Ellis McTaggart, *The Nature of Existence* (1927), V. 2. Cambridge: Cambridge University Press, 1968.

[13] Grünbaum, 'The status of temporal becoming'.

[14] John Leslie, *Immortality Defended.* New Jersey: Wiley-Blackwell, 2007.

[15] Some philosophers of religion and physics argue, against Einstein, that there is absolute cosmic simultaneity, which allows for the perspective of God. This is not broadly accepted. Cf. William Lane Craig and Quentin Smith (eds), *Einstein, Relativity, and Absolute Simultaneity.* London: Routledge, 2007.

[16] Tor Norretranders, *The User Illusion: Cutting Consciousness Down to Size* (translated by Jonathan Sydenham). New York: Viking, 1999, Chapter 9, 'The Half-Second Delay'. Recent work suggests delay as long as seven seconds. Chun Siong Soon *et al,* 'Unconscious determinants of free decisions in the human brain'. *Nature Neuroscience,* 2008, 11, pp. 543–45.

[17] Benjamin Libet, Elwood W. Wright, Jr., Bertram Feinstein, and Dennis K Pearl 'Subjective referral of the timing for a conscious sensory experience'. *Brain,* 1979, 102, (1), pp. 193–224.

[18] Independent tests measure how long it takes between the time a person becomes conscious of something and the time they report that they are conscious of it.

[19] Libet and some others take these experiments as a challenge to free will, since the moment of conscious decision comes after the brain has already started to perform an action. Libet tries to preserve free will as a veto, thinking that consciousness can at least put a hold on an action even if it is the brain that first initiates it. To my mind, the experiments are too far from the topic of free will. Shaun Gallagher argues that free will needs a longer time frame than the 150–350 ms in question here. Shaun Gallagher, 'Where's the action? Epiphenomenalism and the problem of free will', in Susan Pockett, William P. Banks, and Shaun Gallagher (eds), *Does Consciousness Cause Behavior?* Cambridge, MA: MIT Press, 2006, pp. 109–24.

[20] Experimenters cannot fool the brain by asking subjects to make sudden irrelevant decisions, thereby generating actions like finger movements without there having been a half-second of readiness potential in the cortex (Norretranders, 225). Subjects can be made to make sudden movements, but the readiness potential will have already been there before the sudden move.

[21] Patricia Smith Churchland, 'On the alleged backwards referral of experiences and its relevance to the mind-body problem'. *Philosophy of Science,* (1981), 48, pp. 165–81.

[22] Masataka Watanabe, 'Reward expectancy in primate prefrontal neurons'. *Nature,* 1996, 382, pp. 629–32.

[23] James B. Rowe, 'The Prefrontal cortex: Response selection or maintenance within working memory?'. *Science*, 2000, 288, p. 1656.

[24] Benjamin Libet, 'The experimental evidence for subjective referral of a sensory experience backwards in time: Reply to P. S. Churchland'. *Philosophy of Science*, (1981), 48, pp. 182–97.

[25] Daniel C. Dennett, *Consciousness Explained*, *op. cit.* Chapter 6, s. 3 'Libet's Case of Backwards Referral in Time', and s. 4 'Libet's Claim of Subjective Delay of Consciousness of Intention', pp. 153–67.

[26] Errol Morris, *Fast, Cheap, & Out of Control* (film), 1997.

[27] Steven Strogatz, *Sync: The Emerging Science of Spontaneous Order*. New York: Hyperion Books, 2003. Chapter 3, 'Sleep and the Daily Struggle for Sync', pp. 70–100.

[28] Carl Anderson and Francis L. W. Ratnieks, 'Task partitioning in insect societies. 1. Effect of colony size on queuing delay and colony ergonomic efficiency'. *The American Naturalist*, 1999, 154, pp. 521–35.

[29] Carmen Caradima, *Time Delay Compensation in Teleoperation over the Internet*. M.Sc. thesis in Applied Science in Electrical Engineering, University of Waterloo, 1994.

[30] Stéphane Perreault, Robert J. Vallerand, David Montgomery and Pierre Provencher, 'Coming from behind: On the effect of psychological momentum on sport performance'. *Journal of Sport and Exercise Psychology*, 1998, 20, pp. 421–36.

Chapter 6

[1] Henri Bergson, *Duration and Simultaneity* (translated by Herbert Dingle). Indianapolis, IN: Bobbs-Merrill, 1965.

[2] Using Spinoza's distinction between *natura naturans* and *natura naturata*, we can think of a process as natur*ed*, a series of facts or simultaneous snapshots; or we can think of nature natur*ing* itself, flowing contemporaneously into position. It is the same world either way, but what makes flowing reality possible is that instant and flux alternate.

[3] Henri Bergson, *Matter and Memory* (translated by N. M. Paul and W. B. Palmer). New York: Zone Books, 1991.

[4] Hans-Georg Gadamer, *Truth and Method* (translated by Joel Weinsheimer and Donald G. Marshall). New York: Continuum, 1994, p. 352, citing Aristotle, *Posterior Analytics* II, 19, 99ff.

Chapter 7

[1] Jacques Derrida, *The Politics of Friendship* (translated by George Collins). London: Verso, 2006. 'Aphorism countertime', wherein Romeo and Juliet each arrives too late to see the other die (translated by Nicholas Royle), *Acts of Literature* (edited

by Derek Attridge). New York: Routledge, pp. 414–33. 'Force and signification', in *Writing and Difference* (translated by Alan Bass). New York: Routledge, 2001, pp. 1–38. Also, *Schibboleth pour Paul Celan*, Paris: Galilée, 1986, on the locution 'à la fois'.

[2] In Derrida, *Margins of Philosophy* (translated by Alan Bass). Chicago, IL: University of Chicago Press, 1985, pp. 29–68.

[3] Patrick Lee Miller, *Becoming God: Pure Reason in Early Greek Philosophy*. London: Continuum, 2011, pp. 13–17.

[4] Derrida, *Speech and Phenomena* (translated by David B. Allison). Evanston, IL: Northwestern University Press, 1973.

[5] In Derrida, *Margins of Philosophy*, pp. 1–29.

[6] Daniel Giovannangeli, '*Le retard de la conscience*'. *Revue Internationale de Philosophie*, 1998, 205, pp. 367–75.

[7] Jean-Louis Chrétien, *The Unforgettable and the Unhoped For* (translated by Jeffrey Bloechl). New York: Fordham University Press, 2002, p. 50. Chrétien appeals to Emmanuel Levinas, 'Diachrony and representation', in *Time and the Other* (translated by Richard A. Cohen). Pittsburgh, PA: Duquesne University Press, 1987, pp. 97–120.

[8] Quentin Meillassoux, *After Finitude: An Essay on the Necessity of Contingency* (translated by Ray Brassier). London: Continuum, 2008.

[9] Derrida, 'From restricted to general economy: a Hegelianism without reserve', in *Writing and Difference*, pp. 317–50.

[10] Derrida, *Donner le temps*. Paris: Galilée, 1991, p. 51.

[11] Deleuze, *Bergsonism* (translated by Hugh Tomlinson and Barbara Habberjam). New York: Zone, 1990, p. 55ff.

[12] Deleuze, *Difference and Repetition*, p. 59f.

[13] Lampert, *Deleuze and Guattari's Philosophy of History*, *op. cit.* chapters 2–4.

[14] Deleuze, *Cinema 2* (translated by Hugh Tomlinson and Barbara Habberjam). Minneapolis, MN: University of Minnesota Press, 1989, pp. 107–10.

[15] Sigmund Freud, 'Project for a Scientific Psychology' (translated by James Strachey et al), in *The Standard Edition of the Complete Psychological Works of Sigmund Freud*, V. 1. London: Hogarth Press Limited, 1966, pp. 295–343.

[16] Freud, 'Hysterical phantasies and their relation to bisexuality' (translated under Joan Rivière). *Collected Papers, Volume II*. London: The Hogarth Press, 1957, p. 58.

[17] Jacques Lacan, *The Seminar of Jacques Lacan, Book II* (translated by Sylvana Tomaselli). New York: W. W. Norton and Company, 1991, p. 185.

Chapter 8

[1] Deleuze and Guattari, *Mille Plateaux*, Paris: Éditions de Minuit, 1980. See my *Deleuze and Guattari's Philosophy of History*, chapters 7–9.

[2] For Althusser, structural synchrony is expressed even as its operation is delayed. 'Starting with this concept of necessary delay (*retard*), many difficulties can be clarified'. *Lénine et la philosophy*. Paris: François Maspero, 1969, p. 25.

[3] For Horkheimer, 'cultural lag' occurs when social factors change at different rates of speed. 'Authority and the family', in *Critical Theory: Selected*

Essays (translated by Matthew J. O'Connell and others). London: Continuum, 1992, p. 65.

[4] Étienne Balibar, '*Sur les concepts fondamentaux du matérialisme historique*', in Louis Althusser et al, *Lire le Capital* (1965), Paris: Presses Universitaires de France, 1996, pp. 419–568.

[5] For Samir Amin, unequal development is one long history of backlog and catch-up. *Unequal Development: An Essay on the Social Formations of Peripheral Capitalism* (translated by Brian Pearce). New York, NY: Monthly Review Press, 1977.

[6] Balibar, *The Philosophy of Marx* (1993) (translated by Chris Turner). London: Verso, 1995. Chapter 4 'Truth and Progress: Another Philosophy of History?', pp. 80–110.

[7] Balibar, *Race, Nation, Class: Ambiguous Identities* (1988) (translated by Chris Turner). London: Verso, 1991. Chapter 5: 'The Nation Form: History and Ideology', pp. 86–106.

[8] Balibar, *Masses, Classes, Ideas* (translated by James Swenson). New York: Routledge, 1994.

[9] Deleuze, *Cinema 2*(translated by Hugh Tomlinson and Barbara Habberjam). Minneapolis, MN: University of Minnesota Press, 1989.

[10] Heidegger, 'Only a God can save us': The *Spiegel* interview (1966) (translated by W. Richardson), in *Heidegger: The Man and the Thinker* (edited by Thomas Sheehan). Piscataway, NJ: Transaction Publishers, 2009, pp. 45–67.

[11] Ernst Bloch, *Heritage of Our Times* (1935) (translated by Neville and Stephen Plaice). Berkeley, CA: University of California Press, 1991. 'Non-contemporaneity and contemporaneity, philosophically' pp. 104–16. Rebecca Comay cites this text in her analysis of Hegel's reaction to the French Revolution. Germany saw the French Revolution as the delayed effect of its own Lutheran Revolution. 'German history scans as a dissonant counterpoint of divergent rhythms running along separate tracks' (p. 2). 'Absolute knowing is the exposition of this delay…' (p. 6). *Mourning Sickness: Hegel and the French Revolution.* Stanford, CA: Stanford University Press, 2011.

[12] Balibar, '*Préface*' to Yoshiyuki Sato, *Pouvoir et Résistance: Foucault, Deleuze, Derrida, Althusser.* Paris: L'Harmattan, 2007, p. 8.

[13] Jacques Derrida, *Voyous.* Paris: Galilée, 2003.

[14] Lenin, *The State and Revolution* (1917) (no translator listed). Peking: Foreign Languages Press, 1973, p. 147n.

[15] Reinhard Selten, 'Multi-stage game models and delay supergames'. *Theory and Decision,* 1998, 44, pp. 1–36.

[16] Sartre, *War Diaries: Notebooks from a Phoney War 1939–40* (translated by Quintin Hoare). London: Verso, 1999, pp. 122–27.

[17] Jim Vernon, 'The people have spoken: Derrida, democracy, and reciprocal affirmation'. *Int Stud Phil,* 2002, 34, pp. 115–31.

[18] Antonio Calcagno has done the groundwork for comparing Derrida's and Badiou's political texts on delay. *Badiou and Derrida: Politics, Events and their Time.* London: Continuum, 2007.

[19] Alain Badiou, *Being and Event* (translated by Oliver Feltham). London: Continuum, 2005.

[20] Walter Benjamin, 'Theses on the philosophy of history' (translated by Harry Zohn), in *Illuminations* (edited by Hannah Arendt). New York, NY: Schocken Books, 1969, p. 262.

[21] See Zachary Fraser, 'The law of the subject: Alain Badiou, Luitzen Brouwer and the Kripkean analyses of forcing and the Heyting calculus'. *Cosmos and History: The Journal of Natural and Social Philosophy*, 2006, 2(1–2), pp. 94–133.

[22] Cf. the subsection '*Contre le temps*' in '*La politique comme pensée: l'oeuvre de Sylvain Lazarus*', in *Abrégé de Métapolitique*. Paris: Seuil, 1998, p. 44.

[23] Alain Badiou, *Logics of Worlds* (translated by Alberto Toscano). London: Continuum, 2009.

[24] For an introduction to this form of logic, which Badiou relies on in *Logics of Worlds*, try F. William Lawvere and Stephen H. Schanuel, *Conceptual Mathematics: A First Introduction to Categories*. London: Cambridge University Press, 1997. I owe this reference to Luke Fraser.

[25] Alain Badiou, *The Century* (translated by Alberto Toscano), chapter 9 'Seven Variations: Fourth variation: temporal'. Cambridge: Polity Press, 2007, pp. 105–6.

[26] Guy Debord, *Society of the Spectacle* (translated by Fredy Perlman and Jon Supak). Detroit, MI: Black and Red, 1977.

[27] Paul Virilio, *Open Sky* (translated by Julie Rose). London: Verso, 1997, p. 15.

[28] 'The culture industry endlessly cheats its consumers out of what it endlessly promises. The promissory note of pleasure issued by plot and packaging is indefinitely prolonged: the promise, which actually comprises the entire show, disdainfully intimates that there is nothing more to come, that the diner must be satisfied with reading the menu'. Max Horkheimer and Theodor W. Adorno, *Dialectic of Enlightenment* (translated by Edmund Jephcott). Stanford, CA: Stanford University Press, 2002, p. 111.

[29] Adorno, 'A portrait of Walter Benjamin', in *Prisms* (translated by Samuel and Shierry Weber). Cambridge, MA: MIT Press, 1967, p. 237.

Chapter 9

[1] Guillaume Apollinaire, '*Simultanisme-librettisme*' (1914). In *Les Soirées de Paris*, in *Oeuvres completes*, t. 3. Paris: André Balland and Jacques Lecat, 1996.

[2] Oliver Sacks, *Musicophilia: Tales of Music and the Brain*. New York, NY: Alfred A. Knopf, 2007, pp. 112–19.

[3] Why do people want to listen to repeating choruses and codas? Zuckerkandl says the A-A-B sonata form is like watching Act 1 of a play and then being made to watch it again before Act 2. Zuckerkandl's first explanation is that music is like a quilt with a repeated pattern. But a quilt is a simultaneous art form. The spectator can stand repetition if she is only going to attend to it for a second. But where is the pleasure in spending a lot of time looking at each square? Zuckerkandl's second explanation is that time itself is not repeated. Even if a musical phrase is repeated exactly, the first and second listening experiences are different. Victor Zuckerkandl, *Sound and Symbol: Music and the External World* (translated by Willard R. Trask). Princeton, NJ: Princeton University Press, 1969.

[4] The power of attack and decay is essential to rhythm. Stanislavski prepares actors not just to leave a room, but also to open the door with purpose. Constantin Stanislavski, *An Actor Prepares* (translated by Elizabeth Reynolds Hapgood). New York, NY: Routledge, 1989.

[5] Eugenio Barba and Nicola Savarese, *A Dictionary of Theatre Anthropology: The Secret Art of the Performer.* New York, NY: Routledge, 2005, e.g. p. 216.

[6] Other cultures have more notes in the octave. Microtonal classical music is not uncommon.

[7] Aristotle uses 'rhythm' and 'meter' interchangeably in the *Poetics* (translated by W. Hamilton Fyfe). London: William Heinemann Ltd. (Loeb), 1982, 1147a. For provisos, see Pierre Sauvanet, *Le rhythme grec d'Héraclite à Aristote. Paris: Presses Universitaires de France, 1999.*

[8] Between milieu and chaos, as Deleuze and Guattari say in *A Thousand Plateaus*, 'On the Refrain', p. 313.

[9] Daniel J. Levitin, *This is Your Brain on Music.* New York, NY: Dutton, 2006, p. 169.

[10] *A Thousand Plateaus*, 'Becoming-Intense', p. 297.

[11] Serge Ioan Celibidache. *Le jardin de Celibidache* (film), 1998.

[12] Jerry Seinfeld says that the set-up of a joke is a cliff the audience has to leap off, and the punch line is the landing the audience has to reach. If they are too far apart, or too close, the joke does not get a laugh. *Jerry Seinfeld on Comedy* (CD). Laugh.com, Inc., 2001. The film critic Thierry Jousse explains Takeshi Kitano's comedy by 'the principle of the *slow burn* [English original], i.e. of delayed gag'. *Pendant les travaux, le cinema reste ouvert.* Paris: *Cahiers du Cinéma*, 2003, p. 157.

[13] 'The key to the "Fingertips coins across" routine is ... the proper delay between vanish and appearance. Don't perform them simultaneously, lest the trick becomes nothing more than a race between the spectator's eyes and the performer's hand. Perform the vanish first, pause a beat, and then perform the reappearance. The split second between is filled in by the imagination of the audience'. Joshua Jay, *Destroyers: The Superlative Magic of Troy Hooser.* USA: Murphy's Magic Inc., 2001, p. 75.

[14] Philippe Manoury, '*Le temps lisse*'. Presented at the Colloque Écritures du temps et de l'interaction. Paris: IRCAM, June 13–14, 2006.

[15] Marcel Proust, *Le Temps retrouvé.* Paris: Bibliothèque de la Pléiade, t. 3, 1989, p. 872. Deleuze and Guattari cite this passage in *A Thousand Plateaus*, p. 96.

[16] Bernard Sève, *L'altération musicale.* Paris: Éditions du Seuil, 2002, pp. 270–87.

[17] Elliot Carter, 'The time dimension in music' (1965). in *The Writings of Elliot Carter* (edited by Else Stone and Kurt Stone). Bloomington, IN: Indiana University Press, 1977, pp. 243–7.

[18] Jonathan D. Kramer, '*Concepts postmodernes du temps musical*', in (no editor listed), *Le temps et la forme: Pour une épistémologie de la connaissance musicale*'. Genève: Librairie Droz, 1998, pp. 297–322, 340–4. See also Kramer's *The Time of Music: New Meanings, New Temporalities, New Listening Strategies.* New York, NY: MacMillan Publishing Co., 1988.

[19] Daniel Charles defines 'postmodernity' as non-unified temporality. *La fiction de la postmodernité selon l'ésprit de la musique.* Paris: Presses Universitaires de France, 2001.

[20] Jean-Jacques Nattiez, *Le combat de chronos et d'orphée.* Paris: Christian Bourgeois, 1993, 'Le Passé antérieure', pp. 15–31.

[21] Paul Valéry, *Degas Danse Dessin*. Paris: Gallimard, 1938, p. 176. See also Alain Robbe-Grillet, 'Time and description in fiction today', in *For a New Novel* (translated by Richard Howard). New York, NY: Grove Press, 1965, pp. 143–56.

[22] Theodor Adorno, *Philosophy of New Music* (translated by Robert Hullot-Kentor). Minneapolis, MN: University of Minnesota Press, 2006, p. 59.

[23] George Perle, *Serial Composition and Atonality*. Berkeley, CA: University of California Press, 1991. Chapter 5: 'Simultaneity', pp. 84–110.

[24] Perle thinks this confuses harmonies, but Hyde thinks it is the solution. Martha M. Hyde, *Schoenberg's Twelve-tone Harmony*. Ann Arbor: UMI Research Press, 1982.

[25] 'The simultaneous and the successive are all in reality comprised within one unified space'. Arnold Schoenberg, 'Composition with twelve tones', in *Style and Idea* (translated by Leo Black). Berkeley, CA: University of California Press, 1984, p. 220.

[26] Mark DeVoto finds in one five-measure passage, 'a texture of two three-part canons together with a third incomplete three-part canon, partly with collateral voices, all simultaneously over a pedal point (mm. 115–19)'. 'Alban Berg's Three Pieces for Orchestra, Opus 6: Form and thematic structure, and their Relationship to Wozzeck'. www.tufts.edu/ mdevoto/DreiOrch-English.pdf.

[27] Adorno, 'Berg's discoveries in compositional technique', in *Quasi una fantasia: Essays on Modern Music, op. cit.*

[28] Thus saving twelve-tone music from what Adorno calls a 'chronic shortage of notes'.

[29] People say that contemporary classical music is more intellectual than sensuous. But I think that whereas traditional harmony embodies ideas that the listener knows how to anticipate, music without a tonal centre has to be sensuously received in order to be anything at all.

[30] Pierre Schaeffer, *Traité des objets musicaux*. Paris: Éditions du Seuil, 1966.

[31] The order of notes turns out not to matter much to the musical result.

[32] The overtones of different instruments decay differently over time. Drums have overtones, but in-harmonic ones.

[33] Marco Stroppa, for '*Come Natura di Foglia*' (1997), tapes a fisherman's voice, analyzes the sound, and then alters its temporal 'model'. But here, the temporal variations of the music depend on the voices' temporal structure.

[34] Wikipedia 'Pure Data'.

[35] www.sourceforge.net/projects/iannix.

[36] Étienne Darbellay, '*Les masques du temps*', in [no editor listed], *Le temps et la forme, op. cit.*, pp. 191–209, 305–6.

[37] Stockhausen, '...how time passes...' *op. cit.*

[38] François Decarsin, *La musique, architecture du temps*. Paris: L'Harmattan, 2001, pp. 50–70.

[39] Paul Griffiths, *Modern Music and After: Directions Since 1945*. Oxford: Oxford University Press, 1995, pp. 88–90.

[40] Iannis Xenakis, '*Vers une philosophie de la musique*', in *Musique Architectur*. Belgium: Casterman, 1971, pp. 71–119.

[41] Bernd Alois Zimmermann, '*Intervall und Zeit*' [1957] in *Intervalle und Zeit*. Mainz: B. Schott's Söhne, 1974, pp. 11–14.

[42] Claude Debussy, interview quoted in Jean Barraqué, *Debussy*. Paris: Éditions du Seuil, 1962, pp. 118–19.

[43] Laszlo Moholy-Nagy, *Vision in Motion* (1946). Chicago, IL: Paul Theobald and Company, 1965, p. 12.

[44] Guillaume Apollinaire, '*Simultanisme-librettisme*' (1914). From *Les Soirées de Paris, op. cit.*

[45] Filippo Tomasso Marinetti's play *Simultaneità* stages a bourgeois family going through their day, as well as a woman, whom the family cannot see, going through her own day simultaneously on the same stage during the same scenes. In *Teatro,* V. 2 (edited by Giovanni Calendoli). Rome: Vito Biano Editore, 1960, pp. 305–10.

[46] *Futurism & Dada Reviewed* (CD). England: LTM Publishing, 2000.

[47] For a photograph of the dress, Marilyn Stokstad, *Art History*, V. 2. Upper Saddle River, NJ: Prentice-Hall, 1995, p. 1057.

[48] Gérard Grisey, *Vortex Temporum/Talea*, CD liner notes. Multidisc France, 2001.

[49] Catherine Basset, '*Images du monde et traitement du temps dans le gamelan*', in Lévy, Fabien (ed.), *Les écritures du temps (musique, rythme, etc.).* Paris: L'Harmattan, Ircam, 2001, pp. 93–140.

[50] Pierre Sauvanet, *Le rythme et la raison. T. 1 – Rythmologiques.* Paris: Éditions Kimé, 2000. Chapter 3, 'Phénoménologie', pp. 97–154.

[51] See Hegel's vibratory theory of music. *Hegel's Aesthetic: Lectures on Fine Art* Volume II (translated by T. M. Knox). Oxford: Oxford University Press, 1998, pp. 888–933.

[52] José Manuel Berenguer, 'Time, duration, rhythm, consciousness, evolution, attractors, coupled oscillators'. *Actes V: Le temps en Musique Electroacoustique. Bourges: Académie International de Musique Electroacoustique*, 1999–2000, pp. 29–35.

[53] This use of 'plasticity' is suggested by Catherine Malabou (ed.), *Plasticité.* Paris: Éditions Léo Scheer, 2000.

[54] Küpper, Leo, '*Le temps audio-numérique*', in *Ibid*, pp. 94–115.

[55] Curtis Roads, 'The time domain', in *Ibid*, pp. 153–60.

[56] Philippe Schoeller, '*Mutation de l'écriture:Éclat, Stria, DÉsintégrations*', in *InHarmoniques 1: Le temps des mutations*, 1986, pp. 197–208.

[57] Georgiades Thrasybulos, *Music and Language: The Rise of Western Music as Exemplified in Settings of the Mass* (translated by Marie Louise Göllner). Cambridge: Cambridge University Press, 1982.

[58] Theodor Adorno, *In Search of Wagner* (translated by Rodney Livingstone). Manchester: NLB, 1981, p. 79.

[59] Moholy-Nagy, Laszlo, *Painting Photography Film* (translated by Janet Seligman). Cambridge: MIT Press 1969, pp. 41–43.

[60] For interviews with jazz improvisers, see Paul F. Berliner, *Thinking in Jazz: The Infinite Art of Improvisation.* Chicago, IL: The University of Chicago Press, 1994, pp. 289–445. See issues of the journal *Critical Studies in Improvisation/Études critiques en improvisation.*

[61] Frédéric Belivacqua, 'Paradigmes d'interactivité musicale a l'IRCAM'. Presented at the Colloque Écritures du temps et de l'interaction. Paris: IRCAM, June 13–14, 2006.

[62] Adorno, *In Search of Wagner*, p. 99.

[63] Adorno and Hanns Eisler, *Composing for the Films.* London: Continuum, 2007.

[64] Wolfgang Rihm, '*Mutation (Une digression)*'. *InHarmoniques 1: Le temps des mutations,* 1986, pp. 172–80.

[65] Pascal Quignard, *La haine de la musique.* Paris: Éditions Gallimard, 1996, p. 33.

[66] Roland Barthes, *S/Z.* Paris: Éditions du Seuil, 1970, p. 83.

[67] Deleuze, '*Philosophie de la Série Noire*' in *L'Ile Déserte et autres texts.* Paris: Les Éditions de Minuit, 2002, p. 114.

[68] Arnold Schoenberg, 'Composition with twelve tones (2)', in *Style and Idea,* p. 246.

[69] Pierre Boulez, *Boulez on Music Today* (translated by Susan Bradshaw and Richard Rodney Bennett). Cambridge: Harvard University Press, 1971.

[70] Sauvanet, *op. cit.*

[71] Boulez, '*Le systeme et l'idée*'. *InHarmoniques 1: Le temps des mutations,* 1986, pp. 62–104.

[72] Jérôme Baillet, '*Flèches du temps et processus dans les musiques après 1965*', in Lévy, *op. cit.*, pp. 155–215.

[73] Costin Cazabin, '*Temps musical/éspace musical comme fonctions logiques*', in Dufourt, Hughues, Joël-Marie Fauquet, and François Hurand (eds), *L'esprit de la musique: essays d'esthétique et de philosophie.* Paris: Klinksieck, 1992, pp. 301–14.

[74] *Op. cit.*

[75] Hughues Dufourt, *The watery star... L'espace aux ombres* [CD liner notes]. Multidisc France, Accord, 1997.

[76] Bernard Stiegler, '*Programmes de l'improbable, court-circuits de l'inoui*'. *InHarmoniques, op. cit.*, pp. 126–59, written before he became Director of IRCAM.

[77] Edward Dimendberg, *Film Noir and the Spaces of Modernity.* Cambridge: Harvard University Press, 2004, chapter 5, 'Simultaneity, the Media Environment, and the End of Film Noir', pp. 207–59.

[78] Paul Virilio, *Open Sky* (translated by Julie Rose). London: Verso, 1997, pp. 15–20.

[79] Antoine Hervé, '*Temoignages: L'instrument, le mime, le miroir*'. *Instruments* (edited by Peter Szendy), *les cahiers de L'IRCAM.* Paris: Éditions IRCAM, 1995, pp. 89–93, esp. p. 91.

[80] Adorno and Eisler, *op. cit.*, p. 74.

[81] Stiegler, *La technique et le temps, t. 2.* Paris: Éditions Galilée, 1998, pp. 29–30.

[82] Hugo Münsterberg, *The Photoplay: A Psychological Study.* New York, NY: D. Appleton, 1916. Excerpt in Phillip Lopate (ed.), *American Movie Critics: An Anthology from the Silents Until Now.* New York, NY: The Library of America, 2006, pp. 10–17.

[83] Marcel Duchamp, *Duchamp du signe* (notes written in 1912–15 for '*La mariée mise à nue par ses célibataires, même*', published as *La Boîte verte,* 1934). Paris: Flammarion, 1975.

[84] See Jean Suquet, *Miroir de la mariée.* France: Flammarion, 1974.

[85] Alfred Jarry, *Exploits and Opinions of Dr. Faustroll, Pataphysician* (translated by Simon Watson Taylor). Boston, MA: Exact Change, 1996, p. 21.

Chapter 10

[1] John Cage, *A Year From Monday.* Middletown, CT: Wesleyan University Press, 1969, p. 4.

[2] Bertrand Russell, *Human Knowledge: Its Scope and Limits.* New York, NY: Simon and Schuster, 1948, p. 330.

Bibliography

Adorno, Theodor, 'Berg's discoveries in compositional technique'. In *Quasi una fantasia: Essays on Modern Music* (translated by Rodney Livingstone). London: Verso, 1992, pp. 179–200.

—. *Philosophy of New Music* (translated by Robert Hullot-Kentor). Minneapolis, MN: University of Minnesota Press, 2006.

—. *Minima Moralia* (translated by E. F. N. Jeffcott). London: Verso, 1978.

—. 'A portrait of Walter Benjamin'. In *Prisms* (translated by Samuel and Shierry Weber). Cambridge, MA: MIT Press, 1967, pp. 229–41.

—. *In Search of Wagner* (translated by Rodney Livingstone). Manchester: NLB, 1981.

Adorno, Theodor and Hanns Eisler, *Composing for the Films*. London: Continuum, 2007.

Agamben, Giorgio, *Le temps qui reste* (translated by Judith Revel). Paris: Éditions Payot & Rivages, 2000.

Alliez, Éric, *Capital Times* (translated by Georges Van Den Abbeele). Minneapolis, MN: University of Minnesota Press, 1996.

Althusser, Louis, *Lénine et la philosophy*. Paris: François Maspero, 1969.

Amin, Samir, *Unequal Development: An Essay on the Social Formations of Peripheral Capitalism* (translated by Brian Pearce). New York, NY: Monthly Review Press, 1977.

Anderson, Carl and Francis L. W. Ratnieks, 'Task partitioning in insect societies. 1. Effect of colony size on queuing delay and colony ergonomic efficiency'. *The American Naturalist*, 1999, 154, pp. 521–35.

Apollinaire, Guillaume, '*Simultanisme-librettisme*' (1914). In *Les Soirées de Paris*, in *Oeuvres completes*, t. 3. Paris: André Balland and Jacques Lecat, 1996, pp. 888–92.

Aristotle, *Physics* (translated by R. P. Hardie and R. K. Gaye). *The Basic Works of Aristotle* (edited by Richard McKeon). New York, NY: Random House, 1941, pp. 218–394.

—. *Nicomachean Ethics* (translated by Martin Ostwald). Englewood Cliffs, NJ: Prentice Hall, 1962.

—. *Poetics* (translated by W. Hamilton Fyfe). The Loeb Classical Library. Cambridge, MA: Harvard University Press, 1982.

Augé, Marc, *Le temps en ruines*. Paris: Éditions Galilée, 2003.

Bachelard, Gaston, 'The philosophic dialectic of the concepts of relativity'. In *Phenomenology and the Natural Sciences* (edited by Joseph Kockelmans and Theodore Kisiel). Chicago, IL: Northwestern University Press, 1970, pp. 571–2.

—. *L'intuition de l'instant* (1931). Paris: Éditions Stock, 1992.

Bachmann, Ingeborg, *Simultan* (translated by Mary Fran Gilbert as *Three Paths to the Lake: Stories*). New York, NY: Holmes and Meier, 1997.

Badiou, Alain, *Being and Event* (translated by Oliver Feltham). London: Continuum, 2005.

—. *Number and Numbers* (translated by Robin Mackey). Cambridge, MA: Polity Press, 2008.

—. '*La politique comme pensée: l'oeuvre de Sylvain Lazarus*'. In *Abrégé de Métapolitique.* Paris: Seuil, 1998, pp. 35–66.

—. *The Century* (translated by Alberto Toscano). Cambridge, MA: Polity Press, 2007.

—. *Logics of Worlds* (translated by Alberto Toscano). London: Continuum, 2009.

Baillet, Jérôme, '*Flèches du temps et processus dans les musiques après 1965*'. In *Actes V: Le temps en Musique Electroacoustique. Bourges: Académie International de Musique Electroacoustique,* 1999–2000, pp. 155–215.

Balibar, Étienne, '*Sur les concepts fondamentaux du matérialisme historique*'. In Louis Althusser et al, *Lire le Capital* (1965). Paris: Presses Universitaires de France, 1996, pp. 419–568. *Reading Capital* (translated by Ben Brewster). London: Verso, 1979.

—. *Race, Nation, Class: Ambiguous Identities* (1988) (translated by Chris Turner). London: Verso, 1991.

—. *Masses, Classes, Ideas* (translated by James Swenson). New York, NY: Routledge, 1994.

—. *The Philosophy of Marx* (1993) (translated by Chris Turner). London: Verso, 1995.

—. '*Préface*' to Yoshiyuki Sato, *Pouvoir et Résistance: Foucault, Deleuze, Derrida, Althusser.* Paris: L'Harmattan, 2007.

Barba, Eugenio and Nicola Savarese, *A Dictionary of Theatre Anthropology: The Secret Art of the Performer.* New York, NY: Routledge, 2005.

Bardon, Adrian, 'Kant and the conventionality of simultaneity'. *British Journal for the History of Philosophy,* 2010, 18, pp. 845–56.

Barraqué, Jean, *Debussy.* Paris: Éditions du Seuil, 1962.

Barthes, Roland, *S/Z.* Paris: Éditions du Seuil, 1970.

Bassett, Catherine, '*Images du monde et traitement du temps dans le gamelan*'. In *Les écritures du temps (musique, rythme, etc.)* (edited by Fabien Lévy). Paris: L'Harmattan, Ircam, 2001, pp. 93–140.

Baudrillard, Jean, *The Illusion of the End* (translated by Chris Turner). Stanford, CA: Stanford University Press, 1972.

Baugh, Bruce, *French Hegel, From Surrealism to Postmodernism.* New York, NY: Routedge, 2003.

Becker, Oskar, 'Contributions towards the phenomenological foundation of geometry and its physical applications'. In *Phenomenology and the Natural Sciences* (edited by Joseph Kockelmans and Theodore Kisiel). Chicago, IL: Northwestern University Press, 1970, pp. 117–41.

Beierwaltes, Werner, *Platonismus und Idealismus.* Frankfort, KY: V. Klostermann, 1972.

Belivacqua, Frédéric, '*Paradigmes d'interactivité musicale a l'IRCAM*'. Presented at the *Colloque Écritures du temps et de l'interaction.* Paris: IRCAM, June 13–14, 2006.

Benjamin, Walter, 'Theses on the philosophy of history' (translated by Harry Zohn). In *Illuminations* (edited by Hannah Arendt). New York, NY: Schocken Books, 1969, pp. 253–67.

Berenguer, José Manuel, 'Time, duration, rhythm, consciousness, evolution, attractors, coupled oscillators'. *Actes V: Le temps en Musique Electroacoustique. Bourges: Académie International de Musique Electroacoustique*, 1999–2000, pp. 29–35.

Bergson, Henri, *Matière et mémoire*. Paris: Presses Universitaires de France, 1959. *Matter and Memory* (translated by N. M. Paul and W. B. Palmer). New York, NY: Zone Books, 1991.

—. *Duration and Simultaneity* (translated by Herbert Dingle). Indianapolis, IN: Bobbs-Merrill, 1965.

Berliner, Paul F., *Thinking in Jazz: The Infinite Art of Improvisation*. Chicago, IL: The University of Chicago Press, 1994.

Bloch, Ernst, *Heritage of Our Times* (1935) (translated by Neville and Stephen Plaice). Berkeley, CA: University of California Press, 1991.

Boulez, Pierre, *Boulez on Music Today* (translated by Susan Bradshaw and Richard Rodney Bennett). Cambridge, MA: Harvard University Press, 1971.

—. '*Le systeme et l'idée". InHarmoniques 1: Le temps des mutations*, 1986, pp. 62–104.

Breton, André, *Les Vases communicants*. Paris: Gallimard, 1955.

Brown, Blain, *Motion Picture and Video Lighting* (Second edition). Burlington, MA: Focal Press, 2008.

Cage, John, *A Year From Monday*. Middletown, CT: Wesleyan University Press, 1969.

Calcagno, Antonio, *Badiou and Derrida: Politics, Events and their Time*. London: Continuum, 2007.

Caradima, Carmen, *Time Delay Compensation in Teleoperation over the Internet*. M.Sc. thesis in Applied Science in Electrical Engineering, University of Waterloo, 1994.

Carlson, Thomas A., 'Consuming desire's deferral: A theological shadow in the culture of image'. *Parallax*, 1999, 5, pp. 39–55.

Carter, Elliot, 'The time dimension in music' (1965). In *The Writings of Elliot Carter* (edited by Else Stone and Kurt Stone). Bloomington, IN: Indiana University Press, 1977, pp. 243–7.

Cazabin, Costin, '*Temps musical/éspace musical comme fonctions logiques*'. In *L'esprit de la musique: essays d'esthétique et de philosophie* (edited by Hughues Dufourt, Joël-Marie Fauquet, and François Hurand). Paris: Klinksieck, 1992, pp. 301–14.

Celibadeche, Serge Ioan, (director), *Le jardin de Celibidache* (film), 1998.

Charles, Daniel, *La fiction de la postmodernité selon l'ésprit de la musique*. Paris: Presses Universitaires de France, 2001.

Chrétien, Jean-Louis, *The Unforgettable and the Unhoped For* (translated by Jeffrey Bloechl). New York, NY: Fordham University Press, 2002.

Churchland, Patricia Smith, 'On the alleged backwards referral of experiences and its relevance to the mind-body problem'. *Philosophy of Science*, 1981, 48, pp. 165–81.

Comay, Rebecca, *Mourning Sickness: Hegel and the French Revolution*. Stanford, CA: Stanford University Press, 2011.

Craig, William Lane and Quentin Smith, (eds), *Einstein, Relativity, and Absolute Simultaneity*. London: Routledge, 2007.

Darbellay, Étienne, '*Les masques du temps*'. In (no editor listed), *Le temps et la forme: Pour une épistémologie de la connaissance musicale*'. Genève: Librairie Droz, 1998, pp. 191–209, 305–6.

Dastur, Françoise, *Dire le temps*. Fougères: Encre Marine, 1994.
De Grazia, Margreta, 'Teleology, delay, and the "old mole"', *Shakespeare Quarterly*, 1999, 50, pp. 251–67.
De Voto, Mark, 'Alban Berg's Three Pieces for Orchestra, Opus 6: Form and thematic structure, and their relationship to Wozzeck'. www.tufts.edu/~mdevoto/DreiOrch-English.pdf.
Debord, Guy, *Society of the Spectacle* (translated by Fredy Perlman and Jon Supak). Detroit, MI: Black and Red, 1977.
Decarsin, François, *La musique, architecture du temps*. Paris: L'Harmattan, 2001
Deleuze, Gilles, *Proust et les signes*. Paris: Quadrige/Presses Universitaires de France, 1996.
—. *Bergsonism* (translated by Hugh Tomlinson and Barbara Habberjam). New York, NY: Zone, 1990.
—. *Logique du sens*. Paris: Éditions de Minuit, 1969.
—. *Différence et répétition*. Paris: Presses Universitaires de France, 1968.
—. *Cinema 1* (translated by Hugh Tomlinson and Barbara Habberjam). Minneapolis, MN: University of Minnesota Press, 1986.
—. *Cinema 2* (translated by Hugh Tomlinson and Barbara Habberjam). Minneapolis, MN: University of Minnesota Press, 1989.
—. '*Philosophie de la Série Noire*'. In *L'Ile Déserte et autres texts*. Paris: Les Éditions de Minuit, 2002
Deleuze, Gilles and Félix Guattari, *Mille Plateaux*. Paris: Éditions de Minuit, 1980.
Dennett, Daniel, *Consciousness Explained*. Boston, MA: Little, Brown, 1991.
Derrida, Jacques, 'Force and signification'. In *Writing and Difference* (translated by Alan Bass). New York, NY: Routledge, 2001, pp. 1–38.
—. 'From restricted to general economy: a Hegelianism without reserve'. In *Writing and Difference* (translated by Alan Bass). New York, NY: Routledge, 2001, pp. 317–50.
—. '*Différance*'. In *Margins of Philosophy* (translated by Alan Bass). Chicago, IL: University of Chicago Press, 1985, pp. 1–29.
—. 'Ousia and Gramme'. In *Margins of Philosophy* (translated by Alan Bass). Chicago, IL: University of Chicago Press, 1985, pp. 29–68.
—. *Speech and Phenomena* (translated by David B. Allison). Evanston, IL: Northwestern University Press, 1973.
—. *Donner le temps*. Paris: Galilée, 1991.
—. 'Aphorism countertime' (translated by Nicholas Royle). In *Acts of Literature* (edited by Derek Attridge). New York, NY: Routledge, 1992.
—. *Schibboleth pour Paul Celan*. Paris: Galilée, 1986.
—. *The Politics of Friendship* (translated by George Collins). London: Verso, 2006.
—. *Voyous*. Paris: Galilée, 2003.
—. *Peine de mort* (edited by Geoffrey Bennington, Marc Crépon, and Thomas Dutoit). Paris: Galilée, 2012.
Didi-Huberman, Georges, *Devant le temps: Histoire de l'art et anachronisme des images*. Paris: Éditions de Minuit, 2000.
Dimendberg, Edward, *Film Noir and the Spaces of Modernity*. Cambridge, MA: Harvard University Press, 2004.
Duchamp, Marcel, *Duchamp du signe* (published as '*La Boîte verte*', 1934). Paris: Flammarion, 1975.

Dufourt, Hughues, *The watery star... L'espace aux ombres* (CD liner notes). Multidisc France, Accord, 1997.

Einstein, Albert, *Relativity* (translated by Robert Lawson). New York, NY: Three Rivers Press, 1961.

Fano, Michel, '*Les images et les sons*'. *InHarmoniques 1: Le temps des mutations*, 1986, pp. 182–87.

Feynman, Richard P., *QED: The Strange Theory of Light and Matter*. Princeton, NJ: Princeton University Press, 2006.

Fichte, Johann Gottlieb, *Science of Knowledge* (translated by Peter Heath and John Lachs). New York, NY: Appleton-Century-Crofts, 1970.

—. *Science of Rights* (translated by A. E. Kroeger). London: Routledge and Kegan Paul, 1970.

Fraser, Zachary, 'The Law of the subject: Alain Badiou, Luitzen Brouwer and the Kripkean analyses of forcing and the Heyting calculus'. *Cosmos and History: The Journal of Natural and Social Philosophy*, 2006, 1–2, pp. 94–133.

Freud, Sigmund, 'Project for a Scientific Psychology' (translated by James Strachey et al). In The Standard Edition of the Complete Psychological Works of Sigmund Freud, V. 1. London: Hogarth Press Limited, 1966, pp. 295–343.

—. 'Hysterical phantasies and their relation to bisexuality' (translated under Joan Rivière). *Collected Papers, V. II*. London: The Hogarth Press, 1957.

Gadamer, Hans-Georg, *Truth and Method* (translated by Joel Weinsheimer and Donald G. Marshall). New York, NY: Continuum, 1994.

—. 'Plato's unwritten dialectic'. In *Dialogue and Dialectic* (translated by P. Christopher Smith). New Haven, CT: Yale University Press, 1980.

Gallagher, Shaun, 'Where's the action? Epiphenomenalism and the problem of free will'. In Susan Pockett, William P. Banks, and Shaun Gallagher (eds). *Does Consciousness Cause Behavior?* Cambridge, MA: MIT Press, 2006, pp. 109–24.

Giovannangeli, Daniel, '*Le retard de la conscience*'. *Revue Internationale de Philosophie*, 1998, 205, pp. 367–75.

Goethe, Johann Wolfgang von, 'Upon the Laocoon' (translated by Samuel Grey Ward). In *Goethe's Literary Essays* (edited by J. E. Springard). Whitefish, MT: Kessinger Publishing, 2007.

Griffiths, Paul, *Modern Music and After: Directions Since 1945*. Oxford: Oxford University Press, 1995.

Grisey, Gérard, *Vortex Temporum/Talea*, CD liner notes. Multidisc France, 2001.

Groenwegen-Frankfort, Henrietta A., *Arrest and Movement: Space and Time in the Art of the Ancient Near East*. Cambridge, MA: Harvard University Press, 1987.

Grünbaum, Adolf, 'The status of temporal becoming'. In *The Philosophy of Time: A Collection of Essays* (edited by Richard M. Gale). London: Macmillan, 1968, pp. 322–54.

Harman, Graham, *Guerrilla Metaphysics: Phenomenology and the Carpentry of Things*. Chicago, IL: Open Court Publishing Company, 2005.

Hegel, Georg Wilhelm Friedrich, *Hegel's Philosophy of Nature* (translated by Michael John Petry). London: Allen and Unwin, 1970.

—. *Hegel's Aesthetic: Lectures on Fine Art* Volume II (translated by T. M. Knox). Oxford: Oxford University Press, 1998.

Heidegger, Martin, *Being and Time* (translated by John Macquarrie and Edward Robinson). New York, NY: Harper and Row, 1962.
—. *The Fundamental Concepts of Metaphysics* (translated by William McNeill and Nicholas Walker). Bloomington, IN: Indiana University Press, 1995.
—. '"Only a God can save us": The *Spiegel* interview (1966)' (translated by W. Richardson). In *Heidegger: The Man and the Thinker* (edited by Thomas Sheehan). Piscataway, NJ: Transaction Publishers, 2009, pp. 45–67.
Hervé, Antoine, '*Temoignages: L'instrument, le mime, le miroir*'. *Instruments* (edited by Peter Szendy), *Les cahiers de L'IRCAM*. Paris: Éditions IRCAM, 1995, pp. 89–93.
Horkheimer, Max, 'Authority and the family'. In *Critical Theory: Selected Essays* (translated by Matthew J. O'Connell and others). London: Continuum, 1992, pp. 47–128.
Houle, Karen, 'Emendation, *or* when have we been'. In *Lyric Ecology: An Appreciation of the Work of Jan Zwicky*, edited by Mark Dickinson and Clare Goulet. Toronto: Cormorant Press, 2010, pp. 219–30.
Hume, David, *A Treatise of Human Nature* (edited by L. A. Selby-Bigge). London: Oxford University Press, 1973.
Husserl, Edmund, *Zur Phänomenologie des Inneren Zeitbewusstseins. Husserliana X.* Haag: Martinus Nijhoff, 1966.
—. *Cartesian Meditations* (translated by Dorion Cairns). The Hague: Martinus Nijhoff, 1973.
Hyde, Martha M., *Schoenberg's Twelve-tone Harmony*. Ann Arbor, MI: UMI Research Press, 1982.
Jameson, Fredric, *The Prison-House of Language*. Princeton, NJ: Princeton University Press, 1972.
Jammer, Max, *Concepts of Simultaneity: From Antiquity to Einstein and Beyond*. Baltimore, MD: The Johns Hopkins University Press, 2006.
Jarry, Alfred, *Exploits and Opinions of Dr. Faustroll, Pataphysician* (translated by Simon Watson Taylor). Boston, MA: Exact Change, 1996.
Jay, Joshua, *Destroyers: The Superlative Magic of Troy Hooser*. USA: Murphy's Magic Inc., 2001.
Jousse, Thierry, *Pendant les travaux, le cinema reste ouvert*. Paris: *Cahiers du Cinéma*, 2003.
Kant, Immanuel, *Critique of Pure Reason* (translated by Werner S. Pluhar). Indianapolis, IN: Hackett Publishing Company, 1996.
Keller, Pierre, *Husserl and Heidegger on Human Experience*. Cambridge, MA: Cambridge University Press, 1999.
Kierkegaard, Soren, *Either/Or* V. 1 (translated by David F. Swenson). New York, NY: Anchor Books, 1959.
Kockelmans, Joseph, *A First Introduction to Husserl's Phenomenology*. Pittsburgh, PA: Duquesne University Press, 1967.
Koolhaas, Rem and Bruce Mau, *S, M, L, XL*. New York, NY: The Monacelli Press, 1995.
Kosso, Peter, *Appearance and Reality: An Introduction to the Philosophy of Physics*. New York, NY: Oxford University Press, 1998.
Kramer, Jonathan, '*Concepts postmodernes du temps musical*'. In *Le temps et la forme: Pour une épistémologie de la connaissance musicale*' (no editor listed). Genève: Librairie Droz, 1998, pp. 297–322, 340–4.

—. The Time of Music: New Meanings, New Temporalities, New Listening Strategies. MacMillan Publishing Co., 1988.

Küpper, Leo, '*Le temps audio-numérique*'. In *Actes V: Le temps en Musique Electroacoustique. Bourges: Académie International de Musique Electroacoustique*, 1999–2000, pp. 94–115.

Lacan, Jacques, *The Seminar of Jacques Lacan, Book II* (translated by Sylvana Tomaselli). New York, NY: W. W. Norton and Company, 1991.

Ladrière, Jean, '*Raison et eschatology*'. *Revue de l'Université d'Ottawa/University of Ottawa Quarterly*, 1985, 55, pp. 173–91.

Lampert, Jay, *Deleuze and Guattari's Philosophy of History*. London: Continuum, 2006.

—. 'Origen on time'. *Laval théologique et philosophique*, 1996, 52, (3), pp. 649–64.

—. 'Locke, Fichte and Hegel on the right to property'. In Michael Baur and John Russon (eds), *Hegel and the Tradition*. Toronto: University of Toronto Press, 1997, pp. 40–73.

Lawvere, F. William and Stephen H. Schanuel, *Conceptual Mathematics: A First Introduction to Categories*. London: Cambridge University Press, 1997.

Lenin, Vladimir Il'ich, *Materialism and Empirio-Criticism: Critical Comments on a Reactionary Philosophy* (translator not named). New York, NY: International Publishers Co., 1972.

—. *The State and Revolution* (no translator listed). Peking: Foreign Languages Press, 1973.

Leslie, John, *Immortality Defended*. Hoboken, NJ: Wiley-Blackwell, 2007.

Lessing, Gotthold Ephraim, *Laocoon: An Essay on the Limits of Painting and Poetry* (translated by Edward Allen McCormick). Baltimore, MD: The Johns Hopkins University Press, 1962.

—. *Hamburg Dramaturgy* (translated by Victor Lange). New York, NY: Dover Publications, 1962.

—. 'On how the Ancients represented death' (translated by E. C. Beasley and Hellen Zimmern). In *Laokoon and How the Ancients Represented Death*. London: G. Bell and Sons, Ltd., 1914.

Levinas, Emmanuel, 'Diachrony and representation'. In *Time and the Other* (translated by Richard A. Cohen). Pittsburgh, PA: Duquesne University Press, 1987, pp. 97–120.

Levitin, Daniel J., *This is Your Brain on Music*. New York, NY: Dutton, 2006.

Libet, Benjamin, 'Subjective referral of the timing for a conscious sensory experience', co-authored with Elwood W. Wright, Jr., Bertram Feinstein, and Dennis K Pearl. *Brain*, 1979, 102, (1), pp. 193–224.

—. 'The experimental evidence for subjective referral of a sensory experience backwards in time: Reply to P. S. Churchland'. *Philosophy of Science*, 1981, 48, pp. 182–97.

Mackie, Penelope, 'Causing, delaying, and hastening: Do rains cause fires?' *Mind*, 1992, 101, (403), pp. 483–500.

Malabou, Catherine, (ed.), *Plasticité*. Paris: Éditions Léo Scheer, 2000.

Manoury, Philippe, '*Le temps lisse*'. Presented at the *Colloque Écritures du temps et de l'interaction*. Paris: IRCAM, June 13–14, 2006.

Marinetti, Filippo Tomasso, '*Simultaneità*'. In *Teatro*, V. 2 (edited by Giovanni Calendoli). Rome: Vito Biano Editore, 1960, pp. 305–10.

Maritain, Jacques, '*De la métaphysique des physiciens, ou de la simultanéité selon Einstein*'. *Oeuvres de Jacques Maritain*, Volume 3. Paris: Éditions Saint-Paul, 1984, pp. 237–301.

McTaggart, John McTaggart Ellis, *The Nature of Existence* (1927), V. 2. Cambridge: Cambridge University Press, 1968.

Meillassoux, Quentin, *After Finitude: An Essay on the Necessity of Contingency* (translated by Ray Brassier). London: Continuum, 2008.

Merleau-Ponty, Maurice, *Phénomenologie de la perception*. Paris: Éditions Gallimard, 1945.

Miller, Izchak, *Husserl, Perception, and Temporal Awareness*. Cambridge, MA: MIT Press, 1984.

Miller, Patrick Lee, *Becoming God: Pure Reason in Early Greek Philosophy*. London: Continuum, 2011.

Moholy-Nagy, Laszlo, *Vision in Motion* (1946). Chicago, IL: Paul Theobald and Company, 1965.

—. *Painting Photography Film* (translated by Janet Seligman). Cambridge, MA: MIT Press 1969.

Morris, Errol, *Fast, Cheap, & Out of Control* (film), 1997.

Münsterberg, Hugo, *The Photoplay: A Psychological Study*. Excerpt in *American Movie Critics: An Anthology from the Silents Until Now* (edited by Phillip Lopate). New York, NY: The Library of America, 2006, pp. 10–17.

Nattiez, Jean-Jacques, *Le combat de chronos et d'orphée*. Paris: Christian Bourgeois, 1993.

Norretranders, Tor, *The User Illusion: Cutting Consciousness Down to Size* (translated by Jonathan Sydenham). New York, NY: Viking, 1999.

Nowotny, Helga, *Time* (translated by Neville Plaice). Cambridge: Polity Press and Blackwell Publishers Ltd., 1994.

Parker, Patricia, 'Dilation and delay: Renaissance matrices'. *Poetics Today*, 1984, 5, pp. 514–35.

Perle, George, *Serial Composition and Atonality*. Berkeley, CA: University of California Press, 1991.

Perrault, Stéphane, Robert J. Vallerand, David Montgomery and Pierre Provencher, 'Coming from behind: On the effect of psychological momentum on sport performance'. *Journal of Sport and Exercise Psychology*, 1998, 20, pp. 421–36.

Petkov, Vesselin, 'Simultaneity, conventionality and existence'. *TheBritish Journal for the Philosphy of Science*, 1989, 40, pp. 69–76.

Plato, *Timaeus* (translated by R. G. Bury). The Loeb Classical Library. Cambridge, MA: Harvard University Press, 1981.

—. *Philebus* (translated by R. Hackforth). In *The Collected Dialogues of Plato* (edited by Edith Hamilton and Huntington Cairns). Princeton, NJ: Princeton University Press, 1973, pp. 1086–150.

Plotinus, 'On eternity and time' (translated by A. H. Armstrong). The Loeb Classical Library. Cambridge, MA: Harvard University Press, 1967.

Porro, Pasquale, 'Angelic measures: *Aevum* and discrete time'. In Porro (ed.), *The Medieval Concepts of Time*. Leiden: Brill, 2001.

Proust, Marcel, *Le Temps retrouvé*. Paris: Bibliotèque de la Pléiade, t. 3, 1989.

Quignard, Pascal, *La haine de la musique*. Paris: Éditions Gallimard, 1996.

Reichenbach, Hans, *The Philosophy of Space and Time* (translated by Maria Reichenbach and John Freund). New York, NY: Dover Publications Inc., 1958

Richards, Howard, 'Deference'. *Ethics*, 1964, 74, (2), pp. 135–42.

Rihm, Wolfgang, '*Mutation (une digression)*'. *InHarmoniques 1: Le temps des mutations*, 1986, pp. 172–80.

Roads, Curtis, 'The time domain'. In *Actes V: Le temps en Musique Electroacoustique. Bourges: Académie International de Musique Electroacoustique*, 1999–2000, pp. 153–60.

Robbe-Grillet, Alain, 'Time and description in fiction today'. In *For a New Novel* (translated by Richard Howard). New York, NY: Grove Press, 1965, pp. 143–56.

Robinson, Jason, 'Simultaneity and delay: Lampert's already and not yet of now'. In Jay Lampert and Jason Robinson (eds), *Philosophical Apprenticeships: Contemporary Continental Philosophy in Canada*. Ottawa: University of Ottawa Press, 2009, pp. 1–11.

Rosenberg, Daniel and Anthony Grafton, *Cartographies of Time: A History of the Timeline*. New York, NY: Princeton Architectural Press, 2010.

Rowe, James B., 'The prefrontal cortex: response selection or maintenance within working memory?' *Science*, 2000, 288, 1656.

Russell, Bertrand, *Human Knowledge: Its Scope and Limits*. New York, NY: Simon and Schuster, 1948.

Sacks, Oliver, *Musicophilia: Tales of Music and the Brain*. New York, NY: Alfred A. Knopf, 2007.

Sartre, Jean-Paul, *War Diaries: Notebooks from a Phoney War 1939–40* (translated by Quintin Hoare). London: Verso, 1999.

—. '*La fin de la guerre*'. In *Situations, III: lendemains de guerre*. Paris: Éditions Gallimard, 1949.

Sauvanet, Pierre, *Le rythme et la raison. T. 1 – Rythmologiques*. Paris: Éditions Kimé, 2000.

—. *Le rhythme grec d'Héraclite à Aristote. Paris: Presses Universitaires de France*, 1999.

Schaeffer, Pierre, *Traité des objets musicaux*. Paris: Éditions du Seuil, 1966.

Schelling, Friedrich, *System of Transcendental Idealism (1800)* (translated by Peter Heath). Charlottesville, VA: University Press of Virginia, 1978.

Schoeller, Philippe, '*Mutation de l'écriture:Éclat, Stria, Désintégrations*'. In *InHarmoniques 1: Le temps des mutations*, 1986, pp. 197–208.

Schoenberg, Arnold, 'Composition with twelve tones'. In *Style and Idea* (translated by Leo Black). Berkeley, CA: University of California Press, 1984, pp. 214–44.

Seinfeld, Jerry, *Seinfeld on Comedy* (CD). Laugh.com, Inc., 2001.

Selten, Reinhard, 'Multi-stage game models and delay supergames'. *Theory and Decision*, 1998, 44, pp. 1–36.

Sève, Bernard, *L'altération musicale*. Paris: Édition du Seuil, 2002.

Shell, Marc, *Money, Language, and Thought: Literary and Philosophic Economies from the Medieval to the Modern Era*. Baltimore, MD: The Johns Hopkins University Press, 1993.

Sokolowski, Robert, *Husserlian Meditations*. Evanston, IL: Northwestern University Press, 1974.

Soon, Chun Siong, Marcel Brass, Hans-Jochen Heinze and John-Dylan Haynes, 'Unconscious determinants of free decisions in the human brain.' *Nature Neuroscience*, 2008, 11, pp. 543–45.

Stanislavski, Constantin, *An Actor Prepares* (translated by Elizabeth Reynolds Hapgood). New York, NY: Routledge, 1989.

Stiegler, Bernard, *La technique et le temps, t. 2*. Paris: Éditions Galilée, 1998.

—. '*Programmes de l'improbable, court-circuits de l'inoui*'. *InHarmoniques 1: Le temps des mutations*, 1986, pp. 126–59.

Stockhausen, Karlheinz, '....how time passes.....' *Die Reihe*, 1956, pp. 10–40.

Stokstad, Marylin, *Art History*, V. 2. Upper Saddle River, NJ: Prentice-Hall, 1995.

Strogatz, Steven, *Sync: The Emerging Science of Spontaneous Order*. New York, NY: Hyperion Books, 2003.

Suquet, Jean, *Miroir de la mariée*. France: Flammarion, 1974.

Thrasybulos, Georgiades, *Music and Language: The Rise of Western Music as Exemplified in Settings of the Mass* (translated by Marie Louise Göllner). Cambridge, MA: Cambridge University Press, 1982.

Tzara, Tristan, Richard Huelsenbeck and Marcel Janco, '*L'amiral Cherche Une Maison à Louer*' (1916). In *Futurism & Dada Reviewed* (CD). England: LTM Publishing, 2000.

Valéry, Paul, *Degas Danse Dessin*. Paris: Gallimard, 1938.

Vernon, Jim, 'The people have spoken: Derrida, democracy, and reciprocal affirmation'. *International Studies in Philosophy*, 2002, 34, pp. 115–31.

Virilio, Paul, *Open Sky* (translated by Julie Rose). London: Verso, 1997.

Waldenfels, Bernard, 'Time lag: motifs for a phenomenology of the experience of time'. *Research in Phenomenology*, 2000, 30, (1), pp. 107–19.

Watanabe, Masataka, 'Reward expectancy in primate prefrontal neurons'. *Nature*, 1996, 382, pp. 629–32.

Whitehead, Alfred North, *Science and the Modern World*. New York, NY: The Macmillan Company, 1925.

Winckelmann, Johann Joachim, *Reflections on the Imitation of Greek Works in Painting and Sculpture* (translated by Elfriede Heyer and Roger C. Newton). La Salle, IL: Open Court, 1987.

Wundt, Wilhelm, *Outlines of Psychology* (translated by Charles Hubbard Judd). London: Williams and Norgate, 1902.

Xenakis, Iannis, '*Vers une philosophie de la musique*'. In *Musique Architectur*. Belgium: Casterman, 1971, pp. 71–19.

Zimmermann, Bernd Alois, '*Intervall und Zeit*'. In *Intervalle und Zeit*. Mainz: B. Schott's Söhne, 1974, pp. 11–14.

Zuckerkandl, Victor, *Sound and Symbol: Music and the External World* (translated by Willard R. Trask). Princeton, NJ: Princeton University Press, 1969.

Index

Adorno, Theodor 7, 193–4, 201–3, 213–14, 220–1, 236n. 18, 243nn. 28–9
Agamben, Giorgio 65–6
Althusser, Louis 181, 241n. 2
anachronism 2, 180, 183, 193–4, 233n. 1
anticipation 7–8, 13–35, 40–7, 60–1, 77, 92, 150, 154–9, 163, 165, 168, 197–8, 203, 207, 215, 220
Apollinaire, Guillaume 7, 208–9
Aristotle 63–5, 71–3, 144–5, 150–2

Bachelard, Gaston 106
Badiou, Alain 184, 186, 189–95
Balibar, Étienne 171–84
Baudrillard, Jean 13
Baugh, Bruce 236n. 14
Benjamin, Walter 190–1, 194
Berg, Alban 202–3
Bergson, Henri 1–2, 5, 11, 39, 55, 123–47, 148–50, 159–60, 165, 188, 210, 230
Bloch, Ernst 183
Boulez, Pierre 199, 203, 215–19

Calcagno, Antonio 242n. 18
causality 13, 57, 64, 70, 75–82, 89, 106, 167–9, 172, 176, 178, 181, 190, 205, 216, 220, 230, 235, 237n. 6
Chrétien, Jean-Louis 155
Churchland, Patricia 112, 117–18
cinema 95, 163–6, 182–3, 213–14
Comay, Rebecca 242n. 11

Delaunay, Sonia 209
delayed gratification 12, 15–16, 49, 57–63, 115, 157
delayed reaction 12, 14–15, 30, 79, 113, 153–9, 162, 167–71, 182, 230, 235n. 10
Deleuze, Gilles 1–2, 11, 86, 149–50, 155, 159–72, 175, 180–4, 198, 214–19, 234nn. 20, 22, 235n. 10
democracy to-come 184–5, 188, 191
Dennett, Daniel 118–20
Derrida, Jacques 1–2, 18, 35, 66, 149–59, 184–5, 188–91, 195
diachrony 7, 171–2, 177, 179, 182, 184, 211–12
Duchamp, Marcel 223–5

Einstein, Albert 11, 97–106, 133–7
eternity 36, 39–41, 49–58, 62, 65–6, 70, 87, 97, 112, 200

Fichte, Johann Gottlieb 92–3
Freud, Sigmund 155, 167–9
futurism 208–9

Gadamer, Hans-Georg 4, 144
Gallagher, Shaun 239n. 19
Goethe, Johann Wolfgang von 94–6
Grisey, Gérard 204, 209–10
Grünbaum, Adolf 5, 110–11

Hegel, G. W. F. 8–10, 82–96, 131, 146, 211, 234n. 20, 242n. 11
Heidegger, Martin 3, 35, 41–7, 163, 183, 185, 188
history 8–9, 12–13, 22–3, 36–42, 65–6, 72, 79, 91, 94–7, 131, 147, 154–5, 164, 169, 171–84, 189–93, 213, 222–3, 230, 238nn. 12, 18, 242nn. 5, 11
Horkheimer, Max 193, 241n. 3, 243n. 28
Hume, David 78–9
Husserl, Edmund 1–2, 5–6, 8–10, 17–18, 21–35, 43, 46–7, 55, 83, 92, 96, 102, 133, 149–62, 199, 207, 210, 234nn. 1, 2, 18, 235nn. 3–6

Institute for Research and Coordination in Acoustics and Music (IRCAM) 204–5, 218
intersubjectivity 14, 87, 127, 167, 169, 211

Jammer, Max 97, 104–6

Kant, Immanuel 22, 36–9, 75–84, 89–90, 92–4, 235n. 5
Kierkegaard, Soren 32–3

Lessing, Gotthold Ephraim 7, 93–6
Libet, Benjamin 112–18, 239nn. 17, 19

McTaggart, J. M. E. 5, 68, 97, 106–11
memory 11, 14, 18, 30–8, 43–5, 61, 70–3, 87, 111, 117–20, 137–46, 158–60, 164, 168, 211, 215–16, 222–3, 230
Merleau-Ponty, Maurice 35–41
messianic 65–6, 188
Miller, Patrick Lee 151
Moholy-Nagy, Laszlo 208, 213
Murail, Tristan 204, 216, 220

Norretranders, Tor 112–16
now 22–38, 63–6, 86–8, 117–19, 150–8, 194
number 50–1, 53, 63–5, 116, 130, 144, 158, 191–2, 210, 229–30

painting 95, 131, 218, 223–5
past, present, future 2, 5, 12, 14, 17–18, 22–3, 31–9, 42–3, 46, 56, 61–6, 72–5, 82, 87, 102, 106–12, 139–41, 152–68, 194, 216
Plato 49–65, 69
Plotinus 49, 53–6
Porro, Pasquale 56

realtime 193, 212–13, 222–3
Reichenbach, Hans 102–5
retroaction 29–30, 117, 125, 165–8, 174–6, 181, 189–94
return 52–5, 59–65, 69–70, 75–7, 89–92, 101–5, 142, 150, 153, 168, 197, 200–3, 211, 215–16
rhythm 9, 13–14, 19, 22–3, 26–9, 41, 91–2, 116, 120–1, 131–3, 137, 146–7, 173, 177, 198–203, 207–14, 217–19, 223, 230, 244nn. 4, 7

Sartre, Jean-Paul 13, 188
Sauvanet, Pierre 210–11, 217
Schelling, Friedrich 93
Schoenberg, Arnold 7, 199–203, 215–16
sculpture 7, 94–5
sheets of the past 164–5, 230
smooth time 130, 203, 116–23
space 4–7, 10–13, 39, 78–92, 103, 106, 111, 124–32, 139–40, 188, 234n. 20
Stiegler, Bernard 222–3
Stockhausen, Karlheinz 7, 199, 203, 206–7, 218
Strogatz, Steven 120
suspense 220–2
synchronization 6, 12, 16–17, 26, 49–52, 97, 116–20, 134, 199, 214–15
synchrony 7, 104–5, 171–7, 182, 184, 211–12

teleology 57, 63, 66
tempo 15, 26–30, 40–1, 77, 198–201, 206–8, 216–17
theology 15–16, 56, 65–71, 97, 112, 155, 185, 239
time-arrow 13–14, 18, 90–1, 203, 211–12, 215–16, 219–22
time-lines 3, 22–3, 29–30, 34–5, 38, 63–4, 83–5, 110, 138–9, 158–61, 186–8, 194, 210
totum simul 5, 79, 106, 110–12, 140, 194

Virilio, Paul 193, 222–3

waiting 13, 16, 41–6, 85–8, 115, 149, 160–1, 164, 188, 210, 215
Waldenfels, Bernard 42
Wundt, Wilhelm 28

Zuckerkandl, Victor 198, 243n. 3

CPSIA information can be obtained at www.ICGtesting.com
Printed in the USA
LVOW082121081012

302056LV00005B/107/P

9 781441 126399